# Anthropology in the Mining Industry

Glynn Cochrane

# Anthropology in the Mining Industry

## Community Relations after Bougainville's Civil War

Glynn Cochrane
University of Queensland
Brisbane, Queensland, Australia

ISBN 978-3-319-84371-1        ISBN 978-3-319-50310-3   (eBook)
DOI 10.1007/978-3-319-50310-3

Cover design by Samantha Johnson

Printed on acid-free paper

This Palgrave Macmillan imprint is published by Springer Nature
The registered company is Springer International Publishing AG
The registered company address is: Gewerbestrasse 11, 6330 Cham, Switzerland

*For Alison*

# AUTHOR INFORMATION

After five years in the British Solomon Islands Protectorate as a district officer and a district commissioner, Glynn Cochrane completed his DPhil in social anthropology at Oxford University in 1968 and then taught at the Maxwell Graduate School of Syracuse University where he became a department chair and a tenured full Professor of Anthropology and Public Administration. He was the first Director of a Cooperative Agreement between the Maxwell School and the United States Agency for International Development (USAID) on Local Revenue Administration, which conducted research in Asia, Africa, and Latin America.

In 1973 Cochrane wrote a report at the request of the World Bank as to how anthropology could be used in overseas lending operations. Those recommendations were accepted and resulted in the first anthropologists being hired by the Bank. In 1974 he made recommendations on the use of anthropology and wrote *Social Soundness Analysis* (SSA), for USAID. A number of anthropologists were hired and SSA is still in use. He also helped to develop USAID's Development Studies course.

In the 1980s and 1990s Cochrane lived and worked in Papua New Guinea as World Bank Advisor on Public Service reform; in the Cook Islands as United Nations Development Program (UNDP) Advisor to the Prime Minister on Public Sector Reform; in Sri Lanka as Director of a USAID private sector development project; and in Tanzania as World Bank/UNDP Chief Advisor for Civil Service Reform.

In 1995 Sir Robert Wilson, CEO of Rio Tinto plc, asked Glynn Cochrane to make suggestions for the development of community relations for the company. He worked on implementing his recommendations until his retirement in 2002 but maintained contact with Rio Tinto until 2015.

Glynn Cochrane's publications include: *Big Men and Cargo Cults, Development Anthropology, The Cultural Appraisal of Development Projects,* and an edited volume of interdisciplinary studies, *What We Can Do for Each Other.* In 2008 he published *Festival Elephants and the Myth of Global Poverty.* His World Bank publications include: *Policies for Third World Local Government, Reforming National Institutions for Economic Development,* and, with Bernard Liese and Paramjit Sachdeva, *The Organization and Management of Tropical Diseases.*

He is an adjunct professor at the Department of Anthropology and an adjunct professor at the Center for Social Responsibility in Mining at The University of Queensland.

# Acknowledgments

I am grateful to Sir Robert Wilson for providing the opportunity to spend time in the mining industry and to the late Lord (Richard) Holme for his imaginative support. Initially, I worked in London with John Hughes, John Senior, Brian Burgess, George Littlewood, Alexis Fernandez, and Preston Chiaro, and later with Natascia Lillywhite, Simon Wake and Andrew Vickerman; in Argentina with Nicolas di Boscio and Carolyn McCommon; in Indonesia, at Diavik with Doug Willy and Rod Davey; in Australia with Paul Wand, Tim Duncan, and Bruce Harvey; in Indonesia, at Kelian with Mark Hunter, Angus Green, Ramanie Kunanayagam, and Geraldine McGuire; in Indonesian Papua with my late friend Stan Batey, Kal Mueller, as well as Ramanie Kunanayagam, Jeff Tygesen, Diana Wilson and Judy Brown; in Brazil and Peru with Sergio Visconti, Sharon Flynn, David Salisbury, and Louis Cononelos; in East Africa, Zimbabwe, Namibia, and Madagascar with Carolyn McCommon, Marc Demmer, Frank Webster, David Godfrey, Len LeRoux, and Lisa Dean; in Mongolia with Ramanie Kunanayagam and Nicolas di Boscio; in Papua New Guinea (PNG) with Ron Brew; in India with Charles Lutyens and Siddarth Jain; in Borneo with Alan Irving and Angus Green; in Guinea with Peter Adamson, Janina Gawler, and Gregory Maher. Mining industry colleagues included Lizia Lu in Panama, Gary Nash, and Jim Cooney from the World Bank's tri-partite initiative with mining companies, Meg Taylor, an old friend from PNG days, Peter Van Der Veen of the World Bank, and Dr Michele Fulcher of Newcrest Mining Company. While Professor Richard Perry, Dr Catherine MacDonald, Professor Bob Tonkinson, Ron Brew, and John Hughes made constructive suggestions on drafts of the book as did colleagues at the University of Queensland, Professor Deanna Kemp, Professor David Brereton, Professor Saleem Ali, Dr Robin Evans, Phil Clarke, and Dr Diana Arbelaez-Ruiz, responsibility for what is written remains with me.

# CONTENTS

# LIST OF ABBREVIATIONS

| | |
|---|---|
| AAM | Anglesey Aluminium Metal |
| ABG | Autonomous Bougainville Government |
| AGM | Annual General Meeting |
| BCL | Bougainville Copper Limited |
| BHP | Broken Hill Proprietary Limited |
| BIA | Bureau of Indian Affairs |
| BRA | Bougainville Revolutionary Army |
| BSPC | Bougainville Special Political Committee |
| CEO | Chief Executive Officer |
| CERES | Coalition for Environmentally Responsible Economies |
| CESR | Covenant of Economic, Social and Cultural Rights |
| CLO | Community Liaison Officer |
| COP | Communication on Progress |
| COW | Contract of Work |
| CRA | Conzinc Rio Tinto Australia |
| CSR | Corporate Social Responsibility |
| DDMI | Diavik Diamond Mines Incorporated |
| EEZ | Exclusive Economic Zone |
| EPA | Environmental Protection Agency |
| ERM | Environmental Resources Management |
| FAO | Food and Agriculture Organization |
| FFIJD | Fund for Irian Jayan Development |
| FPIC | Free Prior and Informed Consent |
| GDP | Gross domestic product |
| GRI | Global Reporting Initiative |

| GRP | Gross Regional Product |
|-----|------------------------|
| HRAF | Human Relations Area Files |
| IBRD | International Bank for Reconstruction and Development (World Bank) |
| ICMM | International Council for Mining and Metals |
| IFC | International Finance Corporation |
| ILO | International Labor Organization |
| IRMA | Initiative for Responsible Mining Assurance |
| KEM | Kelian Equatorial Mining Company |
| LEMASA | Amunge trust fund |
| LEMASCO | Kamoro trust Fund |
| LKMTL | Association for the Welfare of the Mining Community and Environment |
| MCSC | Mine Closure Steering Committee |
| MD | Medical Doctor |
| MDG | Millennium Development Goals |
| MMSD | Mining and Minerals and Sustainable Development |
| MOU | Memorandum of Understanding |
| NEPA | National Environmental Protection Act |
| NGO | Non-governmental organizations |
| OECD | Organisation for Economic Co-operation and Development |
| OPM | Operasi Papua Merdeka (Free Papua Movement) |
| PNG | Papua New Guinea |
| PNGDF | Papua New Guinea Defence Force |
| PT | Perseroan Terbatas (Limited Liability) |
| PTFI | PT Freeport Indonesia |
| RD | Rain Doctor |
| RPM | Rio Paracatu Mineraçao |
| RPNGC | Royal Papua New Guinea Constabulary |
| RRA | Rapid Rural Appraisal |
| RTZ | Rio Tinto Zinc |
| SDG | Sustainable Development Goals |
| SIA | Social Impact Assessment |
| SLD | Social and Local Development |
| SNSA | Southeastern Natural Sciences Academy |
| SOP | Standard Operating Procedures |
| SSA | *Social Soundness Analysis* |
| STD | Sexually transmitted disease |
| UNDP | United Nations Development Program |

| UNDRIP | UN Declaration on the Rights of Indigenous Peoples |
| UNEP | United Nations Environmental Program |
| UNESCO | United Nations Educational Scientific and Cultural Organization |
| US | United States |
| USAID | United States Agency for International Development |
| WBCSD | World Business Council for Sustainable Development |
| WBG | World Bank Group |

CHAPTER 1

# Introduction

In the 1990s Rio Tinto and other mining companies were beginning to
take stock of how well they were prepared to work overseas. Society's
expectations about what extractive industry should be doing in the community were changing, and this was reflected in a growing tide of suspicion generated by critics of mining. It was becoming clear to mining
companies such as Rio Tinto that they would have to consider changing
the way they did business in the community. I became involved with Rio
Tinto when the CEO of Rio Tinto Sir Robert Wilson, with whom I had
worked in Panama, contacted me when I was the World Bank and United
Nations Development Program's (UNDP) Chief Advisor for Civil Service
Reform in Tanzania. I agreed to provide a report and, after the report was
delivered, was asked to help implement the recommendations.

Mining is not well understood by the general public even in countries
like the USA where the California gold rush helped turn America into an
economic powerhouse. Contemporary large-scale mining may double the
GDP of countries such as Mongolia and Guinea. Mining can also have
negative consequences, such as the illicit trade in alluvial diamonds which
has fueled conflicts in Congo, Liberia, and Sierra Leone. However, industrial diamond operations such as those in Botswana tell a rather different
story. The discovery of diamond deposits transformed Botswana from one
of the poorest countries in Africa to one of the wealthiest. While the country has its share of challenges, such as an extremely high HIV prevalence
of 22 percent, there is no questioning the positive impact of diamonds on
the country's development trajectory. The literacy rate is over 80 percent

© The Author(s) 2017
G. Cochrane, *Anthropology in the Mining Industry*,
DOI 10.1007/978-3-319-50310-3_1

and the per capita income, which tops US$6000, is the highest in Africa. Owing to its high income and educated establishment, the country was the first to provide comprehensive nation-wide anti-retroviral therapy. It is also a leader in biodiversity conservation of rare ecosystems such as the Okavango delta.[1]

New social and environmental attitudes resulted in new regulatory requirements designed to protect communities. A Chatham House conference was told of a recent Goldman Sachs study that showed mining companies could expect to have to make provision for 10–20 times the amount of time that was needed to secure finance, insure, and gain regulatory approval in the 1980s.[2] Indeed, the approval process can become so attenuated that communities lose sight of the real business of the company. In Madagascar Rio Tinto's development of a mineral sands project to produce titanium took so long that local people thought they were dealing with a missionary group. A Citibank study of 400 projects in 2011 showed that two-thirds of the projects were facing substantial delays or cancellation.[3] For a major mining project the cost of every week of delayed production could escalate to US$20 million.[4] Delay in Rio Tinto projects was increasingly due to new regulations and non-governmental organization (NGO) pressures in areas such as land acquisition and resettlement, the rights of indigenous peoples, human rights, environmental protection, labor issues, and benefits for communities.

Of course, longer lead times may also reflect the ability of countries to support mining investments. Countries that are new to mining may have inadequate legal, physical, and financial infrastructure and labor market regulations. For example, in India an executive who had spent many years trying to negotiate a deal for a new mine complained to his headquarters that every tree had its own law. Resource nationalism, the idea that more of the proceeds of mining should remain in the countries who own the minerals, may slow direct foreign investment (FDI) and project development. In such instances the proponents of resource sovereignty say that delay does not matter because the minerals will be safe in the ground and will grow in value. On the other hand, oil companies have done a good job in persuading suspicious governments that they have unique skills that enable them, and only them, to extract oil in a cost-effective way. Mining companies, however, have not done such a good job in persuading suspicious governments that they provide forms of extraction which are more efficient and cost-effective than would be the case if the assets were nationalized.

To raise awareness of the benefits and burdens of mining nine of the world's largest mining companies mounted a research project—Mining and Minerals and Sustainable Development (MMSD)—between 2000 and 2002 to examine how the industry could contribute to the global transition to sustainable development.[5] The new initiative was intended to achieve a serious change in the way industry approached its responsibilities. It included a program of internal reform, a review of the various mining associations to which they belonged, and a rigorous study of the economic and environmental challenges and societal issues they had to face.[6] For example, some of the research concentrated on the problems faced by poor countries when they had windfall profits from mining. The report not only provided a wide audience with information about the mining industry but also ensured that all the top mining executives were familiar with the issues and the possible industry responses.[7] One of the results of MMSD was that the nine founding companies contributed to the establishment of the International Council for Mining and Metals (ICMM) in London.[8]

Managing Directors at Rio Tinto mines overseas were also expected to run all aspects of their operation including community relations. London did not micro-manage but if headquarters did not like the results being achieved then they could, and did, change the Managing Director. For a very big company with mines around the world Rio Tinto was surprisingly decentralized. The corporate headquarters in London was small and with fewer than 200 staff. At first sight these staff seemed anxious to discuss and learn about the functioning of communities. An underlying issue, of course, was the fact that they seemed to respect local communities rather than viewing them as problems to be dealt with.

Miners are proud of what they do and of the contribution their industry makes to society. Large-scale mining is a global industry, with 6000 companies employing 2.5 million people which in developing countries regularly comprises 60–90 percent of total FDI, 30–60 percent of total exports, up to 20 percent of government revenues, and as much as 10 percent of national income.[9] On a practical note the good news is that you can always get hot water in the shower and even hotter water is available for tea lovers because mines have lots of engineers. They do love statistics. Visit a mine office and you may see a plaque celebrating the operation of the plant for a million hours without any loss of time due to injury. The tools of their trade are always on display. There are rocks on shelves that sparkle but are "fool's gold" and not worth anything, core drill samples, geological maps, and exhortations to drink water and watch the color of your urine to avoid dehydration. There will be aerial photos of the mine site with the pit from which ore is being extracted at the bottom. Technical drawings and photographs are more common than artists' renditions. You see rocks in the lobby of a mining office, pictures of gigantic trucks with unbelievably large tires, and photos of what they can do in a collision with a light vehicle. Visitors at a remote mine may think they are in the back of beyond; miners know they are on the front line.[10]

Miners use words like "beneficiation," which means to strengthen metal before smelting, and "dewatering," which means to pump out water from a mine. They are above all practical, pragmatic people who take safety seriously. Miners take their safety knowledge home with them: stuff like working at heights, operating any equipment that can hurt you or your family. A mine is usually an alcohol-free zone, and miners are used to being tested to make sure they have not used alcohol or any illegal substances.

Miners do not use a lot of adjectives. With the exception of exploration you do not hear a great deal about outstanding geologists, exceptional plumbers and electricians, or rather unusual instrumentation expertise. The miner is a unit of work, a box on the organization chart, a space in a dormitory. What mining values in the individual worker is getting the job done without mistakes. Keep your head down and get on with it. While the anthropologist needs a setting where he or she can stand out, the miner is happy to be like others in a community of miners.

Anthropologists are by nature dedicated individualists whose professional success may appear to outsiders to depend in large part on being different, and these differences can come from working in a unique culture, wearing very unusual clothing from that area, or voicing eye-catching opinions gained from fieldwork. When anthropologists visit a mine site they usually want to get to the community right away. However, the engineers have a different idea of what a visitor needs to know. The visitor to a mine site is likely to be told about the amount of power consumed, the liters of water saved and recycled, the tons of rock and soil called overburden that have to be scraped away to let miners excavate the ore. You may be told that the pit can be seen from outer space and that it would take 10 years to fill with water. Even though a visitor may be interested in communities, his or her first stop at site will be to admire the depth of the pit, the skill with which the benching has been done, and so on. In fact, before visiting the mine site there will be a small but important rite of passage where the visitor is separated from others, clothed in helmet, safety boots, safety glasses, and ear plugs, and then, as a lookalike miner, provided with an introductory tour.

Anthropologists like their opinions to stand out and be noticed. Try to imagine the job interview between an executive who is a mining engineer specializing in pit slope stability and a recent anthropology PhD who says that his or her thesis explored the analogy between obstetrics and mining, with ore equated to embryo, mine to uterus, shaft to vagina, and miner to obstetrician (sic).[11] What would the executive make of a report which said: "Mount Kare was traditionally a ritual site where pigs were sacrificed to Taiyundika, a totemic python, to protect the fertility of plant, animal, and human species?" Today, gold is mined in pursuit of unprecedented riches and millenarian transformations. Although sacrifices are no longer conducted at Mount Kare, the python still has some salience for Paielas, who consider the gold to be the flesh of the totemic python. Blending Christianity with traditional cosmology, Paielas interpret the finding of gold as a millenarian sign.[12]

Anthropology is a social science and its practitioners are rightly wary about predicting the future, and in the past when they, inevitably, asked for more studies before giving an opinion this went down badly with engineers. Confidence in the discipline was not helped by the fact that if you put five anthropologists in a row and asked for an opinion you might well get five quite different answers. Anthropologists do not pretend to be more able to predict the future than other social scientists but Delphic utterances are, of course, a temptation. Those who felt they knew a community better than anyone were sometimes willing to take the small but important step from science to prediction by saying they knew what would be best for that community.

What communities think about a development proposal from a company or a government is not easy to determine from a distance. It is unwise to assume that members of the community, as the lowest rung on an administrative ladder that stretches up to the capital, will know their place and be properly respectful. Nor can it be assumed that those who make important development decisions are continually on the lookout for any signs of community feeling that may seep upwards. And, since it is only rarely the case that good statistics are available, it is clear that whoever wants reliable information has no alternative but to establish and maintain regular personal contact with communities themselves.

The community was the intellectual birthplace of anthropology, so community relations provided me with an at-home feeling.[13] Community relationships can carry economic content, political content, and a touch of the sacred.[14] My experience outside the mining industry inclined me to the view that the community should be a basic unit of account for government, aid agencies, and NGOs. It makes sense for those who want to help to ask themselves how well they understand the communities concerned and how reliable are the community views that influence their assistance. I had spent five years in the British Solomon Islands Protectorate slithering up and down muddy mountains in the tropical rain forest, sleeping in remote villages, talking to local people long into dark nights, getting malaria and tropical ulcers whose scars are still with me, and learning a little about the importance of listening and observing carefully. The community provided an arena where government could be confronted by local people who were sometimes supportive and understanding and sometimes angry. Later, after spending time at Oxford studying anthropology and then being engaged in university teaching for a number of years, I entered the service of the World Bank and other UN agencies. I discovered an amazingly broad portfolio of opportunities. Although the focus was on international and national issues my experience had been with local government and communities. As time passed I found it frustrating to be always giving advice without actually seeing the advice through. As Rio Tinto's first full-time professional anthropologist I found the possibility for developing and using hands-on implementation skills as well as having a long-term association with communities; a refreshing change from my aid agency experience.[15]

One of the things that surprised me as I became more familiar with community relations in the mining industry was the certainty that was continually expressed by those who believed in tool kits about what should be done in the community. Getting it right was apparently pretty straightforward. However, the volumes of guidance and advice that were provided and seldom consulted indicated that professionals on the ground shared my reservations about being told that all would become clear after reading some page or other. The need for freedom and opportunity to assume risk increases when the goals sought and the means of accomplishing them become more ambiguous. For example, fairly specific norms of behavior and expectations can be applied to an engineering program; this kind of mission may need relatively little freedom. But when the mission is to improve community livelihoods the ambiguity is very high for two reasons: first, relatively little agreement may exist about what should be done, about the necessary actions, or the required competence or experience; second, we are not dealing with overt, predictable behavior. It is apparent, of course, that operations do not fall easily into the polar positions of high or low predictability and conformity, and low predictability: they tend to be arranged on a continuum between the two extremes. Nevertheless, it is useful to recognize the extremes as a basis for understanding the kind of organization and climate that fosters innovation and improvement.

What I found satisfying about working for Rio Tinto was that I could draw on what I had learned while working for the World Bank and UNDP as well as my time in academia but with some differences. Written opinions for Rio Tinto executives had to be brief and to the point if you wanted support. A three-page draft was likely to generate questions about the discursive style of the author. When I submitted my first substantial company report my boss said, "Professor, the anthropology is interesting. But everyone here is busy. If you cut this down we can get the right people to read it." I knew that when they called you "professor" it meant that you were not really being practical so I made painful cuts. "Look," he said, "you are getting there. Just cut some more and we will get it to the Director." Out went the references and most of everything that I had thought important. "Oh dear," he said looking at the two-page draft report, "this is fine but also pretty much common sense and I'm not sure we need specialists for this kind of thing."

A few companies have explored the idea of outsourcing or contracting out responsibilities for community relations. I remember asking Leon Davis during his time as CEO of Rio Tinto if the community function could be outsourced and he said he did not think so. His view was that the process of getting minerals out of the ground could easily be outsourced but not community relations. "That is who we are," he said.

This book is addressed to those who are interested in learning about the relationships between mining companies and communities as well as relationships between mining companies, NGOs,[16] UN agencies, and regulatory authorities. My focus is on the steps taken by Rio Tinto and other mining companies to improve community relations performance around the world in the light of a growing number of accusations of human rights abuses and cultural insensitivity in the mining industry. I refer to other mining companies when my remarks have relevance to the mining industry. I concentrate on remote communities in developing countries because that is where Rio Tinto has mines, as do other companies. Many of the world's largest deposits of copper and iron ore that remain to be developed are found in the likes of Guinea, Afghanistan, Mongolia, and the Democratic Republic of the Congo.[17] I should stress that, in addition to their influence on performance in developing countries, the community relations improvements that Rio Tinto introduced also resulted in a significant contribution to the company's community performance in industrialized countries.

## ORGANIZATION OF THE BOOK

Part I illustrates ways in which mining companies were accused of being complicit in wrongdoing. I begin with Bougainville and the island's civil war tragedy before turning to mining and its effects on indigenous peoples in Panama and around the world.[18] The difficulties faced by indigenous peoples when they try to preserve or conserve their cultural heritage are acknowledged. Given the need for miners to ensure that land purchase or alienation involving indigenous people meets traditional legal norms as well as the requirements of the state, special attention is paid to land tenure and the difficulties involved in changing group tenure to individual land holding. Part I concludes by examining how and why mining companies joined the UN Global Compact. As an obligation of membership mining companies were obliged to submit regular reports on their environmental and community performance to UN agencies and NGOs.

Part II discusses the business case for community relations and the policy, skills, and organization that were put in place to develop Rio Tinto's systematic approach to community relations. Experience suggested that it was important to listen to communities and learn from them before engagement. This approach consisted of three sequential steps that were designed to reach a deeper level of appreciation of how to work with local societies, beginning with a social baseline. The baseline provided information that was necessary for the second step, which was to reach a deeper level of understanding about how to consult and reach understandings that had local validity. The first two steps laid a foundation for the third, which was to try to understand ways in which communities might be helped to undertake social and economic development in partnership with the company. Artisanal mining and closure, two of the industry's most intractable problems, are discussed showing the ways in which the systematic approach could help with resolution.

Part III looks at corporate social responsibility (CSR), free prior and informed consent (FPIC), and other examples of social responsibility thinking. The CSR concept lacks intellectual rigor as does its use on many occasions. The book suggests that when CSR and FPIC are used without first making an attempt to get to know the community involved the results are likely to be unreliable. CSR does not provide a satisfactory substitute for fieldwork and hands-on community competence. A chapter is devoted to the important issue of resettlement where the social responsibilities of mining companies have frequently been questioned. In looking at the performance of mining companies in the community it is obvious that NGOs and UN agencies headquartered in New York, Washington, London, and Geneva have not been close to communities or local issues. They have tended to work from their offices trying to get the right flagship policies in place with supporting programs and projects; mining companies have pursued hands-on, smoke-in-the-eyes, relationship-building in communities. Although they do not have hands-on community skills, a fieldwork tradition or organization located close to communities, the aid agencies, and big NGOs have continued to maintain that they know what is best for communities that they may not know at all well. Doing the hard yards in the community and demonstrating tick-box compliance with the reporting requests of international agencies and organizations enabled Rio Tinto and other miners to gain a reputation for competence in the community. The book concludes by reviewing the most important factors that have made for the success of hands-on community relations.

## NOTES

1. See Ali (2009).
2. See Chatham House (2013).
3. See Sainsbury et al. (2011).
4. Cited in Davis and Franks (2011).
5. See Wilson (2000).
6. See International Institute for Environment and Sustainable Development (2002).
7. Unfortunately there has been almost a complete change at the top of mining companies and many of today's CEOs were not part of that MMSD process.
8. It is sometimes interesting when there is a lazy tendency to use sustainability as a slogan to push for deeper explanations of the case for and against the sustainability movement. For a non-believer account see Becker (1996).
9. See International Council on Mining and Minerals (2014).
10. See Trigger (1997).
11. See Eliade (1962). For surveys of mining and anthropology see Godoy (1985), Hyndman (1987, 1997), Taussig (1980), Nash (1979).
12. See Biersack (1999).
13. The importance of the ethical community possessing shared values and beliefs was emphasized by Robertson Smith (1889).
14. For an interesting idea of the number of dimensions that can be crammed into a social relationship see Fortes (1963).
15. On my work on the introduction of anthropology to the World Bank see, Goodland, Robert J. (1999), *Social & Environmental Assessment to Promote*

*Sustainability: An Informal View from the World Bank*. Glasgow: International Association of Impact Assessment. On United States Agency for International Development (USAID) and social soundness see Alan Hoben, Alan (1982), "Anthropologists and Development," *Annual Review of Anthropology*, American Anthropologist, Vol. 11, pp. 347–394.

16. NGO, or the umbrella-term "NGO," refers to a broad, kaleidoscopic grouping of organizations that can differ considerably in size, resources, organizational level, mission, form, and orientation. In recent decades, the number of NGOs worldwide has increased dramatically, along with a strengthening of their influence and a broadening of their activities. Some of these organizations are, like aid agencies, better at raising money than spending it. While statistics about global numbers of NGOs are notoriously incomplete, it is currently estimated that there are somewhere between 6000 and 30,000 national NGOs in developing countries. It is now estimated that over 15 percent of total overseas development aid is channeled through NGOs. Much of the interest of the NGOs is focused on poor countries where democratic ways of doing things are either of recent provenance or fragile disposition.

17. See Cattano (2009).

18. See Perry (1996).

## BIBLIOGRAPHY

Ali, Saleem H. 2009. *Treasures of the Earth: Need, Greed and a Sustainable Future*. New Haven: Yale University Press.

Becker, Wilfrid. 1996. *Small is Stupid: Blowing the Whistle on the Greens*. London: Duckworth.

Biersack, Aletta. 1999. The Mount Kare Python and His Gold: Totemism and Ecology in the Papua New Guinea Highlands. *American Anthropologist* 101(1): 68–87.

Cattano, Ben. 2009. *The New Politics of Natural Resources: Time for Extractive Industries to Address Above-Ground Performance*. London: Environmental Resources Management.

Chatham House. 2013. *Revisiting Approaches to Community Relations in Extractive Industries: Old problems, New Avenues?* London: Chatham House.

Davis, R., and D. Franks. 2011. The Cost of Conflict with Local Communities in Extractive Industry. In *Proceedings of the First International Seminar on Social Responsibility in Mining*, eds. D. Brereton, et al., ICMM, Santiago, October 19–21.

Eliade, Mercia. 1962. *The Forge and the Crucible*. London: Rider.

Fortes, Meyer. 1963. Ritual and Office in Tribal Society. In *Essays on the Ritual of Social Relations*, ed. Max Gluckman. Manchester: Manchester University Press.

Godoy, Ricardo. 1985. Mining: Anthropological Perspectives. *American Review of Anthropology* 14: 199–217.

Hyndman, David. 1987. Mining, Modernization, and Movements of Social Protest in Papua New Guinea. *Social Analysis* 21: 20–38.

Hyndman, David C. 1997. The Archaeology and Anthropology of Mining: Social Approaches to an Industrial Past. *Current Anthropology* 38: 20–39.

International Council on Mining and Minerals. 2014. *The Role of Mining in National Economies*. 2nd ed. London: International Council on Mining and Minerals.

International Institute for Environment and Sustainable Development. 2002. *Breaking New Ground: Mining, Minerals and Sustainable Development*. London: International Institute for Environment and Sustainable Development.

Nash, June. 1979. *We Eat the Mines*. New York: Colombia University Press.

Perry, Richard. 1996. *From Time Immemorial: Indigenous Peoples and State Systems*. Austin, TX: University of Texas Press.

Robertson Smith, William. 1889. *Lectures on the Religion of the Semites. Fundamental Institutions*. First Series. London: Adam & Charles Black.

Sainsbury, C., C. Wilkins, D. Haddad, D. Sweeney, et al. 2011. *Generation Next: A Look at Future Greenfield Growth Projects, Citi Investment Analysis*. New York: Citibank.

Taussig, M. 1980. *The Devil and Commodity Fetishism In South America*. Chapel Hill: University of North Carolina Press.

Trigger, David S. 1997. Mining, Landscape and the Culture of Development Ideology in Australia. *Cultural Geographies* 4(2): 161–180.

Wilson, Sir Robert. 2000. *Meeting the Challenge to 21st Century Mining*. Davos: World Economic Forum.

# PART I

# Bougainville

# CHAPTER 2

# Bougainville Lessons for Rio Tinto

When the former Prime Minister of Papua New Guinea (PNG) Michael Somare said in an affidavit given long after the Civil War that Rio Tinto was complicit in the action by PNG armed forces to regain control of the mine, the company vehemently denied the claim. After Bougainville, mining companies faced allegations of environmental damage, human rights abuses, heavy-handed security, ignoring workers' rights, and cultural genocide. As a result, for the last 25 years community relations in the mining industry has been shaped by charges of being complicit in wrongdoing. Mining did indeed contribute to the Bougainville Civil War but it was far from being the primary cause of that conflict, for deep down the islanders wanted to be independent of the state of PNG.[1]

The Bougainville Civil War gave birth to "complicity," a new form of societal wrongdoing. Mining companies accused of complicity began

---

The chapter on Bougainville draws on my time in what was then Western District in the British Solomon Islands Protectorate (BSIP) where as an Administrative Officer I met with my Australian counterparts from PNG from time to time; experiences as a World Bank Advisor resident in Port Moresby in the 1980s; contact and conversations with Douglas Oliver, whom I got to know in the 1960s when he sent his graduate student Roger Keesing to do research on Malaita where I was working in the Solomon Islands; Hugh Laracy in Auckland who is especially knowledgeable about the Catholic Marist missionaries; and Gene Ogan whom I met in Bougainville and in Honolulu.

© The Author(s) 2017
G. Cochrane, *Anthropology in the Mining Industry*,
DOI 10.1007/978-3-319-50310-3_2

by being guilty and continued to be guilty because imagination and suspicion arose in situations where no hard evidence could be produced.[2] Accusers hoped the public would be horrified by what had been alleged, while the accused were confident that the charges could not be backed by credible evidence. Critics began by making the most sensational eye-catching charges they could dream up. Complicity could not be quickly countered because the slow, patient building of community relationships took place out of sight of the public and could not match the speed, power, and front-page impact of campaigns promoted by those who wanted to stop the opening of new mines or to close down existing operations. Naturally, companies were not anxious to begin by acknowledging error or shortfall since there was a good deal at stake: reputation, prestige, resources, credibility, and authority. Complicity charges were a media event, an academic fashion, which produced tough questions for mining companies and the possibility of fundraising for the accusers. Like fireworks, the illumination faded quickly.

The 1988–1998 Bougainville conflict, which is believed to have killed between 15,000 and 20,000 people, was a disaster for PNG and the island of Bougainville. International observers were outraged after women and children died when a naval blockade cut off medical supplies to the island and PNG armed forces were accused of throwing prisoners out of helicopter gunships into the sea. Following the Civil War any mining company that wanted to claim a competent performance in the community faced an uphill battle to gain credibility. Civil War turned Bougainville into a Melanesian melanoma that needed to be excised quickly, which could not happen because the roots of the conflict were deep and historic and intertwined in a number of ways, producing results that had relatively little to do with mining.

In 1964, when PNG was still an Australian territory, geologists found low-grade copper mineralization at Panguna on Bougainville Island where the Nasioi people lived. The mine opened in 1969 but by 1990 this mine, owned by Conzinc Rio Tinto Australia (CRA),[3] and operated by its subsidiary Bougainville Copper Limited (BCL), and which was one of the largest and most profitable copper mines in the world, was closed by local people. Between 1972 and 1989 some 3 million tons of copper and 9.3 million ounces of gold were mined from Panguna. At the time of closure the mine was contributing 40 percent of PNG's annual revenue.[4]

The closure was engineered by Francis Ona, the leader of the rebellion, who had resigned his job at the mine in 1988 and went into the jungle to form the Bougainville Revolutionary Army (BRA). Shortly thereafter Australians employed by BCL were shot at and at least one wounded and owing to the escalating threat of violence and loss of life BCL's mine was shut down on 15 May 1989 and mothballed in 1990. Francis Ona was helped by an Australian-trained lieutenant, Sam Kauona, who had defected from the Papua New Guinea Defense Force (PNGDF) to become Ona's right-hand man. He broke into the mine's magazine, stole dynamite, and blew up electricity pylons. He had been trained by the Australian army in the use of dynamite. The island was placed under the control of the PNG Police Commissioner at the time, Pius Kerepia, who was from Bougainville (he was later shot and killed in Port Moresby by unidentified assailants). In May 1990, PNG imposed a blockade on Bougainville which resulted in many deaths because local people no longer had access to essential medicines.

What was the cause of this violent disagreement? Under the 1974 Agreement between PNG and BCL the people of Bougainville felt that they had to suffer the development of a mine most of the profits from which went to a government they disliked. The price of gold rose in 1974 and again in 1979 without any change to the 1974 benefit-sharing agreement. There had been other irritants. For example, the clumsy manner in which the Australian colonial administration had acquired the land needed for the mine from the traditional landowners in the late 1960s and early 1970s produced a landowner riot that ended up on the front page of Australia's national newspapers.[5] Any help that anthropological land tenure studies might have provided was negated by government's practice of drawing straight lines on map lines; these often did violence to the rights of various lineages and clans whose land did not fit neatly into these blocks.[6] In PNG only a very small amount of land had been registered[7]: most of the land was held under traditional tenure.

While compulsory acquisition was obviously legal in terms of national law it was inevitable that a compulsory land acquisition would be locally misunderstood and bitterly resisted. Instead of recognizing the strength of custom and moving slowly to negotiate and to try to find a peaceful way forward, the Bougainville administration decided to take control of the land they wanted.[8] Local landowners said that what was under the ground belonged to them[9] although the Australian government made it very clear that it considered the revenues from minerals to belong to the state. By late 1988 the level of violence resulted in the deployment of the Royal Papua New Guinea Constabulary (RPNGC) Mobile Squads and elements of the PNGDF. Although initially restricted to the area around the mine site, the conflict subsequently intensified. Abuses were reported to have been committed against the population by both sides during fighting between government forces and the rebels of the BRA with the ensuing conflict developing into a general separatist insurgency. Fighting continued for a year, during which widespread human rights violations were alleged to have occurred, including the burning of many villages. However, in early 1990 PNG withdrew, leaving Bougainville in the control of the BRA. The PNG government subsequently imposed a blockade on Bougainville. Australian support as part of this program included funding for the PNGDF, provision of arms and ammunition, logistics, training, and some specialist and technical advisors and personnel. The blockade remained in effect until the ceasefire in 1994 (although it was informally continued for some parts of Bougainville until 1997).[10]

The island began to descend into disarray. The command structure set up by the BRA seldom had any real control over the various groups throughout the island that claimed to be part of the BRA. A number of *raskol* (criminal) gangs affiliated with the BRA, equipped largely with weapons salvaged from the fighting in World War II, terrorized villages, engaging in murder, rape, and pillage. Bougainville split into several factions as the conflict took on ethnic and separatist characteristics.

When Peter Lawrence published his classic book on cargo cults, *Road B'Long Cargo*[11] he emphasized the fact that Melanesian leaders keep looking for the right road which will lead to the fabulous riches represented by cargo.[12] Francis Ona went one better. After he closed the Bougainville mine as leader of the rebellion, he established a sovereign state called Meekamui, which was composed of communities around the mine, made himself King, issued his own currency, provided law and order, and made it known that he wanted 10 billion kina (about $US3.2 billion) in compensation from BCL. Rumors circulated that money invested in Meekamui could earn 1000 percent profit.[13] Meekamui residents still try to exchange their currency at PNG banks but without success.

Life in Meekamui was far from ideal. The blockade of the island imposed by the national government had resulted in a shortage of medicines and skilled health care, great hardship, and international condemnation—it may also have contributed to Francis Ona's death in 2006 following a bout of malaria. The absence of schools and schooling in the Meekamui state produced a generation of young men and women who were neither literate nor numerate. The Meekamui residents lived above the tree line, which meant that cash crops, cocoa, and copra could not be produced. The environmental situation deteriorated in the 20 years following mine closure with extensive and uncontrolled acid rock drainage. Small amounts of money were earned from the sale of scrap metal salvaged from the mine. Artisanal miners panned for gold in the "blue river," the color of which came from copper tailings.

## Bougainville: A Solomon Island Society

With the benefit of hindsight it is now obvious that the Bougainville Civil War represented a much broader disagreement between Bougainville and PNG than dissatisfaction over the terms of the 1974 Mining Agreement. Bougainville did not want to be a part of PNG and this had been clear long before the mine arrived—the mine closure was simply another incident in a long line of attempts to win independence.

The border between the British Solomon Islands Protectorate (BSIP) and PNG had been set by a British Pacific Order-in-Council in 1893. Bougainville had strong trading and kinship ties with the Shortland and Treasury Islands, Choiseul and Gizo in the Western District of the Solomon Islands. Marriages between the two areas were common, and they both used traditional shell wealth which consisted of fathoms (the term used in the islands) of shells strung together. It was one of the Solomon Islands and had residents who were as black as those in the western Solomons of the British Protectorate to the south. Pius Kerepia, the former Police Commissioner and later the Commissioner for the Corrective Institutions Service who had been shot, was from Bougainville. The Bougainville people were often teased about being so black. John Vulupindi, the PNG Secretary of Finance (from West New Britain), said to Kerepia, "You are so black you could get out of one of your prisons in the night without anyone seeing you."

In 1972 the Bougainville Special Political Committee (BSPC) was set up to negotiate with the Papua New Guinean government on the future of Bougainville within PNG. As a concession to Bougainville, the PNG administration established Provincial government in the early 1970s and the North Solomons Provincial government had its Headquarters at Arawa. On 28 May 1975 the interim provincial government in Bougainville agreed to secede from PNG. This caused a three-way impasse between the government of PNG, the legislature in PNG, and the authorities in Bougainville. The PNG government made attempts to resolve the situation through June and July of that year, but these failed, and the interim government announced that they would declare independence on 1 September, ahead of PNG's own independence on 16 September. On that day, they issued the Unilateral Declaration of Independence of the Republic of North Solomons. The Bougainville Agreement was signed later that year establishing the North Solomons Provincial Government and giving Bougainville widespread autonomy within PNG and independence was promised within five years. This then led very quickly to the decentralization of the PNG political framework, with provincial government being established in all 19 provinces.

In 1975 the author was a member of a World Bank Economic mission to PNG on the eve of independence the purpose being to assess that country's credit-worthiness. The mission had some concerns about the future of the PNG economy, almost half of whose revenues came from the mine, in the event that Bougainville decided to leave PNG. Having been an administrator in the BSIP which bordered Bougainville to the south of the island the author was asked by the mission leader to go to Honiara to quietly ask what the reaction might be in the Protectorate should Bougainville announce an intention to leave PNG. The Chief Secretary, Mr. Wynn-Jones said that such an approach would not be welcomed.

The report produced by the 1975 World Bank mission was never released. Mission members discussed the findings in Canberra with Sir John Crawford of the Australian National University and Gough Whitlam, then Prime Minister of Australia. The mission leader, Gunter Reif, later told me that an Australian Director of Economics at the World Bank had stopped the release of the report though Reif did not know why. Stopping publication suggested to some that Australia was worried about how well PNG had been prepared for independence. But there is another strange and possibly zany possibility. Australia had been given responsibility for British New Guinea in 1905, four years after Australian Federation and this was confirmed by the League of Nations after World War I and by the United Nations after World War II. Australia was, in strict point of fact, administering a colony on behalf of Britain. Was Australia, which was not legally independent until 1986, in a position to give PNG independence in 1975?

## Copra Came Before Copper

For Bougainvilleans mining on the island was not the first time that large sums of money had arrived; money had been earned from producing copra (the white meat of the coconut) and cocoa for half a century before copper was discovered.[14] For 50 years plantations in PNG had been worked by recruited laborers from nearby islands. But the mine's demand for labor, and skilled labor at that, needed a different approach. PNG did not have a sufficient number of skilled tradesmen. The mining company wanted to recruit in Asia. However, the Australian government did not want an influx of foreign workers, particularly workers from Asia. In the end fewer than 2000 Asian workers were given temporary entry. The introduction of workers from elsewhere in PNG, rather than those from Asia, was another major reason for the closure of the mine. Thousands of the workers at the mine were brought in from the New Guinea highlands. The locals called them "redskins" because they were so much lighter in skin color than people from Bougainville. The highlanders were aggressive. They pushed and shoved to get their own way. When they did not succeed, fighting broke out. Men were killed and women raped.[15] As dissatisfaction mounted the Bougainville people took violent action to expel all "redskins" from their island.

## The Importance of Kinship

On Bougainville a group of men claimed by right of seniority that they were the owners of the land to be mined and subsequently, as the Panguna Land Owners Association, they became the recipients of most of the rents and compensation paid by the company. This system was recognized under PNG legislation which entrenched the leadership in place at the time and allowed them to sequester the flow of funds. The Panguna men were acting on behalf of their sisters because women, not men, owned land on Bougainville.[16] Was the money going to the right people? Did leaders of the association, the brothers of the women landowners, give their sisters their fair share of the money they received?

Intergenerational conflict was common enough in Melanesia at the time. To the south in the British Solomon Islands I remember schoolboys called their elders "kumara tops." Near the mine on Bougainville young educated men were furious with their elders, who they thought were being duped. The division between the young and the old and between men who managed their sisters' properties may have contributed to the eventual tragedy on the island. Had the local matrilineal kinship system been affected by patrilineal principles? Such a development although rare in mainly patrilineal PNG had been noticed elsewhere in the country. This possibly rare change in kinship behavior was suggested when cocoa producers on the Gazelle Peninsula of New Ireland were unwilling to sell their beans to a processing plant. Some male producers wanted their sons to inherit their wealth rather than their nephews as was customary under a matrilineal system. This was because the sales would be registered and thus proceeds would be expected to go to a farmer's nephew, which would be expected in line with matrilineal inheritance, instead of to his son as was the case when patrilineal principles were followed.[17]

## WHAT THE CRITICS SAID

Most of the post-conflict comments about Bougainville have been made without any of the authors viewing the situation on the ground because, since the end of the Civil War, the mine had been a "no-go" zone. Analysts and commentators both within and outside the mining industry differed in their views as to the reason for the conflict. For some it was an example of what colonialism had left behind, and for others it was an example of greedy capitalism. There were also those who saw it as an example of the micro-nationalism that bedeviled small Pacific island states. Bougainville and its mining communities were presented as the hapless victim of an unnecessary Civil War. It was said that Panguna residents had not only suffered collateral violence, rape, and death but also livelihood destruction because their agricultural land had been taken and used for mining without their consent; they had not received a fair portion of the wealth from an uncaring government and a complicit company; and their environment had been forever damaged by BCL's unthinking pursuit of profit.

## A RETURN TO MINING?

When the Civil War ended, the Autonomous Bougainville Government (ABG) was created (and is still part of PNG but with a referendum on independence from PNG to be held by 2020). On the island there were 20 distinctive ethnic groups with mutually unintelligible languages and the population, estimated to be growing at 4 percent per annum, had reached an estimated 250,000 in the 2011 census. Bougainville was in desperate need of money for social services and investment; however, with a few small exceptions—the Japanese built a new road from Buka to Arawa—most of the foreign aid had been for law and order projects as donors waited for proof that disarmament had taken place and that peace would be maintained. Although peace on Bougainville had been reestablished with the help of New Zealand and Australia no on-the-ground research had been done in and around the mine for over 20 years because Meekamui had been closed to outsiders.

In 2014 an uneasy peace ceremony was agreed between the ABG and the PNG government which owned 20 percent of the mine. It was marked by a ceremony in which Premier Momis and Prime Minister O'Neill broke arrows to signify hostilities were at an end. While the PNG government, appeared for the moment to have escaped the need to make reparations, the Meekamui rebels were still saying that BCL needed to come up with the old figure of 10 billion kina presented by Francis Ona. Trying to get the company to the table, the ABG suggested a Bel Kol ceremony at which BCL could apologize and make a symbolic payment in respect of those who had died. But Bel Kol simply meant "cooling the belly," something that was needed to calm down protagonists after a tribal fight. It would also be tantamount to a public admission of guilt. Following such an admission, demands for very serious compensation would be made.

The trust deficit between the mining communities and BCL was substantial. Discussion about a return to mining was orchestrated by the John Momis–led Bougainville government. Momis had been deputy prime minister in the PNG government when the mine closed and he sided with the rebels. Later he became PNG Ambassador to China and thereafter was suspected of wanting to bring in Chinese investors to run the mine. To get the Meekamui rebels to even discuss a reopening of the mine and an end to the no-go zone, the ABG had to pass a law in 2015 saying that mineral wealth belonged to the landowners. An earlier attempt to split the mining wealth 50/50 between landowners and the ABG was rejected.

The Meekamui rebels who controlled the mine site did not seem anxious to share the wealth from mining with any other group on the island or elsewhere in PNG. As a consequence any notion that the proceeds of mining were for all citizens ended with Bougainville. The revenue arrangements for subsequent mines in PNG reflected a sense that mining wealth must be seen as local wealth that primarily belonged to the landowners. Because traditional society was egalitarian with access to land for all and had no landless class the emphasis on "landowners" was strange and somewhat un-Melanesian.[18] The idea that the landowner owned minerals under the ground was not clearly established in Melanesian culture.[19]

To the mining industry, and even the critics of the mining industry, the idea of investing several billion dollars into bringing the mine back to production was not a very attractive proposition. Any future investor would have to consider the possibility that a return to mining would require the development of a new method for the disposal of the tailings produced by processing the ore because by 2000 there was general agreement in the mining industry that riverine disposal of tailings should be avoided (see Annex A). In addition, environmental remediation, according to published United Nations Environmental Program (UNEP) reports, could cost a great deal of money, much of which expenditure could have been avoided had Meekamui allowed BCL to access the site in order to continue to treat the tailings to reduce the risks of toxicity while the mine was closed.

The Panguna landowners' negotiating position represented a strong form of what might be called "resource localism," and it continually conveyed a powerful sense that most of the profit mining companies made was not justified. Meekamui rebels were suggesting that, with a little help, local people could do the mining themselves. What if the government of PNG nationalized the mine as it had done with Ok Tedi, the owner of which had been BHP Billiton? Would not the ABG try to get as much as possible out of the mine to fund island-wide development? If workers were brought in from elsewhere how could a repeat of the redskin problem be prevented? What if one or two local hotheads decided that they were not getting enough out of the mine and decided to shut it down?

It was doubtful that there would be many developers who would want the property badly enough to meet the terms local people seemed determined to demand, and determined to be able to alter when community sentiment changed. The landowners wanted to control the profits from mining but were the landowning groups any better constituted than when the mine closed? How much had the matrilineal inheritance system changed in the years since the mine closed? Would women landowners get their fair share? Would violence and intimidation stop? The local skills base had been eroded, so where would workers come from and who would the islanders allow to work at the mine? Bougainvilleans appeared not to like anyone from anywhere else.[20]

## RIO TINTO COMMUNITY RELATIONS

BCL was proud of the fact that the company had built a first-class mine in a remote region which had exceeded all production targets. BCL had done everything the government had told them to do; they believed they had dealt fairly with local people and simply could not understand what had gone wrong. There had been no warning. How could a government lose control? Who in their right mind would want to throw all this achievement away?

After Bougainville, when new mines were started in PNG plumbers, electricians, machine operators, and heavy equipment operators who had been produced by BCL's very successful skills development efforts were easily found. Would events have turned out differently if Australia had originally put a better deal on the table, one that recognized local rights to mineral ownership? This would have helped but probably not enough. Bougainville did not fit well into PNG and would probably continue to have a stronger sense of island affiliation and nationhood than adherence to the State of PNG.

BCL had been generous in its dealings with communities. The company had provided schools, hospitals, and clinics as well as buildings for local government. A charitable foundation, which is still functioning, was started to help local development in agriculture and small-scale enterprise. BCL had personnel in the community, some of whom were local. Their job was to hand out money, and they functioned in much the same ways as PNG's Mining Wardens, who tried to sort out land and boundary disputes, disagreements over Trust Fund payments, and local employment issues. BCL assumed that making agreements and spending lots of money on health, education, and local economic development would ensure good relationships.

Racism was charged by some. It was said by the critics of mining that BCL had no good local understanding and few employees who were known and trusted in the community. But the reality was that without exception outsiders—government, company, migrant workers—had failed to establish relationships based on mutual trust and respect with the people of Bougainville. The outsiders did not usually marry locally or settle locally; the expatriates worked and lived in relative isolation from the local people.

It was true, and obviously truer still as time passed, that Panguna people did not believe that they had anybody they could trust to give them sound advice. They had complained that there was apartheid, with a separation between expatriates and local employees reflected in housing, after-hours socializing, and recreational club memberships for tennis, sailing, and swimming. In the upper echelons of the company managers were keen to understand local society and to plan to try to make improvements. However, at lower levels of the company relationships with local people could be rough and crude. In later years when expatriates who had worked on Bougainville went to work elsewhere in the mining industry they carried with them a reputation as people who "kicked arse ocker style."[21]

Proponents of giving communities a bigger package believed all would be well with future mines if the price was right.[22] Out of this came the realization that in developing countries it was not enough to do a deal in the capital; the deal also had to be done and maintained locally. Instead of social relations some mine managers wanted legal relations, thinking that what was needed was a legal deal with local people. Subsequently those who thought this way failed to appreciate the power of social change and deep community disagreement, which should have been one of the enduring lessons of Bougainville, and so they kept returning to the idea that a deal on paper had the capacity to control local behavior and to get approval in the boardroom.

BCL had engaged Douglas Oliver, a Harvard university academic anthropologist who had worked for many years on Bougainville issues.[23] Although Oliver advised on the design of accommodation for the mine and on likely relationships between local people and the miners, he did not have an opportunity to spend enough time to probe deeper into possible future problems. Oliver stressed in his report to CRA that knowledge of the local language was vital. He was mainly concerned with kinship and linguistics, whereas the key issues at the mine centered on the disposition of the wealth created by the mine and land tenure. The Siwai grew cocoa whereas the coastal Nasioi were copra producers, and it was the production of copra that brought the Nasioi into contact with "redskins" from PNG. The Catholic Marist missionaries,[24] who had links to the southern Solomons, were strong in Nasioi, whereas the Methodists were strong in Siwai. Douglas Oliver's fieldwork was too early to give him personal experience of the effects of the introduction of cash cropping and population increase on Bougainville. This meant that he had limited personal experience of the strength and importance of intergenerational tension or of the pressures for a shift to patrilineal inheritance, both of which have proved to be so difficult for Bougainville and other Melanesian islands. CRA did not know what questions should be put to Oliver, and he did not know a great deal about mining. It was also the case that Oliver was working with Catholic Nasioi people rather than the Methodist Siwai, where he had done his own fieldwork in the late 1930s and 1940s.[25]

Bougainville represented a considerable setback for those who believed that social science could help to create and sustain better relationships between miners and their neighbors. The company had tried to understand local culture but this had not worked; there had been no warning, no identification of the pressure points, and no tactics or strategies for managing disagreement. The real determinants of future community behavior at the mine seemed to be a reflection of continuous and continuing competition among aspiring leaders. In an unsettled situation, jockeying for position was so persistent and outcomes were often so temporary and uncertain that only those living close to the community and in constant touch with the main actors had a chance of understanding what was going on or what was likely to happen. The concerns of outsiders were not the same as internal concerns.[26]

For those anthropologists who wanted to see an improvement in the community relations of the mining industry improved through the use of their science there was a chilling message from the American Anthropological Association. Douglas Oliver told me that he was charged with a violation of the Code of Ethics of the Association because he refused to make public his reports to CRA (the code was promulgated as a result of the involvement of anthropologists with the Vietnam War and with the CIA in Latin America).[27] The American Anthropological Association and Oliver's accusers had established a first: Miners needed watching, because their work could be classified as clandestine research.[28]

## RIO TINTO AND BOUGAINVILLE AFTER THE WAR

During the 25 years that followed the trauma of the war Rio Tinto kept a very low profile in relation to Bougainville. Court cases alleging complicity in starting the war were vigorously, and successfully, defended. London supplied a Managing Director and a board member to BCL, and BCL continued to operate the Bougainville Foundation for the benefit of all Bougainvilleans. What London did not do was position Rio Tinto for the questions that would be asked when, inevitably, there were questions about whether or not the company wanted to reopen the mine. It would have been useful to try to position the company for a reentry into the limelight by making the case for the fact that Rio Tinto had not started the war and doing more to combat the accusation that the company had aided the PNG government in its prosecution of the war.

It probably did not occur to many observers that the PNG government, which had somewhat recklessly imposed a naval blockade and invaded the island thereby causing a great deal of death and suffering, would eventually emerge from the conflict without any need to provide compensation to Bougainville while all the local parties—Panguna, ABG, and the PNG government—would unite in implying that Rio Tinto was the villain.

## ENVIRONMENTAL DEGRADATION

The 1990s produced a number of high-profile environmental incidents— Ok Tedi in PNG, Marcopper in the Philippines, toxic chemicals released near a World Heritage site in Spain, red mud from aluminum smelting reaching the Danube in Hungary—all of which suggested that miners had a rather casual attitude toward environmental protection (Annex A). What irritated NGOs and the public was the fact that the mining industry did not seem to get it; they seemingly could not understand why it was that the public was so concerned about environmental issues.

Since the copper was low grade, over a billion tons of material had to be processed and the residue, or tailings as miners call it, was dumped into the Jaba River, where they wiped out fish stocks; the tailings then flowed into Empress Augusta bay damaging the marine environment. The tailings were also said to be responsible for causing birth defects, and the extinction of the flying fox on the island.[29] Bougainvilleans were angered by the damage to their environment caused by the mine and made worse during the closure because environmental remediation which had accompanied mining was discontinued after the Civil War.

BCL's environmental professionals had been concerned to meet international standards and benchmarks but paid little attention to meeting and protecting local perceptions, which in their eyes lacked scientific credibility.[30] Not enough weight had been given by BCL to local feelings about the pollution of the Jaba River and the elimination of flying foxes, which were a local delicacy.[31]

Harm caused to the environment was so obvious to local people and international observers that it did not need to be the subject of a complicity charge. However, NGOs and the general public began to believe that if a company was careless with its environmental responsibilities it was also more than likely that it was careless with the welfare of neighboring communities.

Miners regarded the disposal of tailings as a technical matter for the company, and for the country where the mine was situated, to decide between them. If the government and the regulators gave approval why should anyone else be involved? Miners made and announced their decision about how they intended to dispose of untreated or treated tailings in a river or by impoundment in a dam on engineering and economic grounds, and then if necessary they defended that decision. It was, as far as the miners were concerned, inconceivable that their technical judgment could be vigorously and noisily challenged. What was not factored into this internal decide-announce-defend calculation about tailings management was the possibility of public outrage and community outcry. This attitude changed when the regulators in industrialized countries began to require bonds up to the total cost of remediation.

After Bougainville NGOs and the public assumed, on occasion correctly, that the environmental behavior of mining companies was pretty much the same thing as their behavior in the community. Bad environmental behavior could easily be detected and measured with some precision so there was an assumption that critics of the mining industry could do the same thing with regard to community activity. As time passed the environmental performance improved markedly and could be measured and assessed from a distance. But complicity charges proved to be sticky because community relations did not lend themselves to metrical standards and quantitative assessment, unlike environment or health and safety remediation.

## An Exit from Bougainville is Announced by Rio Tinto

In 2013 BCL announced the results of an Order of Magnitude study which suggested that there was still substantial value in the company's mining tenements. Based on this survey the minority shareholders in BCL wanted to see a return to mining, and they, together with the ABG, began to put pressure on Rio Tinto to make a decision about a return to mining. The ABG converted BCL's license to mine into exploration licenses. Rio Tinto then announced that the company would conduct a two-year review of its Bougainville investment.

On 30 June 2016 Rio Tinto announced an end to its Bougainville involvement and gifted the 53.8 percent of BCL shares they owned (worth about US$40 million) in equal portions to the ABG and the PNG government. This donation was unlikely to be welcomed by Bougainville and even less by the Panguna landowners. In a public notice Rio Tinto said that the Panguna mine had been operated by BCL in accordance with PNG law and therefore there was no Rio Tinto involvement. BCL had been expelled by Bougainvilleans and was subsequently denied access to the mine site. Rio Tinto did not intend to accept any responsibility for any social and environmental remediation that had been made necessary by the acquisition of mining tenements, resettlement of local people near the mine, or the war and its aftermath. Rio Tinto would no longer continue the practice of providing BCL with a Managing Director.

What Rio Tinto also needed to do, but did not, was to say "sorry" about the suffering and damage caused by the conflict in a public, locally meaningful, way; not in a Bel Kol ceremony which, as already stated earlier, would have implied that Rio Tinto was guilty and culpable, but in a peacemaking ritual signifying joint participation in an expression of regret for the past such as the breaking of arrows (the ritual settlement with the ABG used by the PNG government) accompanied by a feast with many pigs and puddings. After saying "sorry" Rio Tinto could have said publicly that the company was not responsible for all the damage that Bougainville suffered. At that point, the company could, and many would think should, have offered some symbolic assistance, perhaps in coordination with the ongoing, and new, efforts of aid agencies and NGOs, to help the island prepare for a return to mining. Any symbolic assistance might not have cost much because in Melanesia making an effort to achieve reconciliation is what is important rather than how much compensation is involved. When Tom Beanal, the tribal leader of the Amungme, on whose lands Freeport's copper mine in Indonesian Papua is located, sued and lost a US$6 billion lawsuit against Freeport MacMoran, he is reputed to have asked James R. Moffett, Chairman of Freeport MacMoran, if Moffett could buy him a truck.

Although the ABG reacted with fury to the exit it did not look as if the exit would produce a global campaign against Rio Tinto. These days mining companies are more likely to face local campaigns sparked by resource nationalism than traditional NGO global campaigns about human rights, labor issues, or cultural damage. In Indonesia the prominent environmental campaigner *Wahana Lingkungan Hidup Indonesia* (WALHI) had to adopt a lower profile when its campaigns began to be seen as critical of the mining industry decisions of the government; in the UK, Oxfam is seldom on the attack perhaps because that is not the image of a major NGO that wishes to position itself as a serious global player and one that has become a regular recipient of UK government financial support; even Greenpeace has gone quiet.

Resource nationalists want to change the terms of the mining concessions held by foreign miners in their countries. They say that the foreign miners get windfall profits and are allowed to produce unacceptable damage to health and environmental safety, while doing too little to help local communities. Local resource nationalists claim their countries have the skills to do the mining themselves. Miners such as Rio Tinto have been slow to respond effectively. They needed to start by making a case a number of years ago based on the fact that they bring with them technologies and approaches that are locally unavailable, such as unique smelter technology, or unique underground mining expertise. When asked some years ago by the author to say what benefits foreign miners had brought, the Mayor of Antofogasta in Chile, where there are large copper mines, said that foreign miners had successfully introduced the idea of safety and good practice had spread to all businesses in the region. Foreign miners can also show that they have unique skills and experience with community relations and community investment. That is important because sustaining the local and regional acceptability of mining is a neglected, though critically important and critically contentious, issue for resource nationalists who are often out of touch with their rural populations and not equipped to anticipate and address community concerns.

## NOTES

1. See Filer (1990). See also Ogan (1999).
2. Henry VIII's Treason Laws operated in much the same way. See Bellamy (1979).
3. Rio Tinto Zinc (RTZ), a UK company, had a substantial shareholding in CRA which in 1995 merged with RTZ to become Rio Tinto plc, one of the largest mining companies in the world.
4. See Palomka (1990).

5. See Denoon (2000).
6. See Jorgenson (1997).
7. Registration under the 1962 Land Ordinance was clumsy since it required registering ground which would, after registration, be held under primogeniture in the names of six males. Lineages, which might also be matrilineal, had several hundred members. How did one choose six males from a lineage which might have more than a hundred members? See also, Simpson (1976).
8. Eugene Ogan told me in 2000 that the Australian District Commissioner on Bougainville at the time was "a horse's ass" and that independence had been 50 years in the making.
9. See Ogan (1971).
10. See May and Spriggs (1990).
11. Those interested in this common form of cult-thinking could look at one of the following: see Worsley (1964) or Cochrane (1970).
12. See Filer (1990). See also Ogan (1999).
13. Was there any precedent for Kingship? African anthropologists have tongue-in-cheek asked why their leaders were just "chiefs" and not Emperors. In Melanesia the idea of a "Paramount Chief" seems to have crept in from Fiji where there was such a distinction. Roger Keesing summarized some of this creative thinking about the past. See Keesing (1989).
14. See Ogan (1996).
15. See Lasslett (2009). Lasslett adopts a Marxist perspective that does not have a very good fit with Melanesian society. He goes further than Worsley (1964), who also adopted a Marxist position in *The Trumpet Shall Sound* (London, Methuen). Lasslett's problem with the Marxist approach is that Melanesian society does not really have a concept of class that is so essential for Marxist analysis. He would have done better, since he is located in Ireland, to consider the question Irish historians puzzled over as to why Ireland did not lend itself to Marxist analysis. See, "How Marx and Engels Erred When They Applied the Notion of Class to Ireland," in *Marx and Engels on the Irish Question* (Progress Publishers, Moscow, 1974).
16. Local land tenure is shaped by an unusual Dravidian matrilineal system with bilateral cross-cousin marriage and both uxori-local and viri-local residence. See Hage (2004). These systems have been changing rapidly under the impact of the introduction of the cash economy and the extent to which this process has advanced in the area has not been regularly assessed. See Nash (1974).
17. See Epstein (1969), Ogan (1991).
18. This landowner question was one where the Melanesian Alliance political party had the same perspective on land ownership as the Bougainville landowners.

19. See Cochrane (1974).
20. See Filer (1990).
21. Personal communication from an ex-BCL employee.
22. For one example of what might be done, see Skalnik (1989).
23. Douglas Oliver worked for CRA from 1968 till 1974. See Oliver (1955, 1991).
24. See Laracy (1976).
25. For examples of the use of anthropology see Belshaw (1976), Brokensha (1966), Forster (1969), Hoben (1982), Angrosino (1976), Grillo and Rew (1985).
26. See Jackson (1991).
27. I met Douglas Oliver in Honolulu in 2000. Following the dismissal of the ethics charge, Oliver was surprised, in view of the trouble he had over refusing to make his notes public, to find that Denoon (*Getting Under Their Skin*) had permission from the mining company to publish material which he had given them in the form of highly confidential reports.
28. See Dreger (2011).
29. Today, according to UNEP, the damage to the environment is even more obvious than it was 30 years ago because local people would not allow outsiders to visit and remediate as would have happened had the mine continued to operate. As a result environmental controls and procedures to mitigate the toxic impact of the tailings deposition could not be put in place.
30. To get a sense of extreme environmental criticism see West (1972). In its obituary of Richard Wilson on 28 April 2015 *The Times* pointed out that West was successfully sued by Rio Tinto for statements in the book.
31. See Brosius (1999), Kirsch (2015).

## BIBLIOGRAPHY

Angrosino, Michael V., ed. 1976. *Do Applied Anthropologists Apply Anthropology?* Athens: University of Georgia Press.
Bellamy, John. 1979. *The Tudor Law of Treason.* London: Routledge.
Belshaw, Cyril. 1976. *The Sorcerer's Apprentice: An Anthropology of Public Policy.* New York: Pergamon.
Brokensha, David. 1966. *Applied Anthropology in English-Speaking Africa.* Ithaca, NY: Society for Applied Anthropology, Monograph 8.
Brosius, Peter J. 1999. Analysis and Interventions: Anthropological Engagement with Environmentalism. *Current Anthropology* 40(3): 277–310.
Cochrane, Glynn. 1970. *Big Men and Cargo Cults.* Oxford: Oxford University Press.

————. 1974. Land Alienation: The Case for Traditionalists. *Oceania* 45(2): 124–131.

Denoon, Donald. 2000. *Getting Under the Skin: The Bougainville Copper Agreement and the Creation of the Panguna Mine.* Melbourne: Melbourne University Press.

Dreger, Alice. 2011. Darkness's Descent on the American Anthropological Association: A Cautionary Tale. *Human Nature* 22: 225–246.

Epstein, A.L. 1969. *Matupit, Land, Politics and Change Among the Tolai of New Britain.* Berkeley: University of California Press.

Filer, Colin. 1990. The Bougainville Rebellion, The Mining Industry and the Process of Social Disintegration in Papua New Guinea. *Canberra Anthropology* 13(1): 1–39.

Forster, George M. 1969. *Applied Anthropology.* Boston: Little Brown.

Grillo, Ralph, and Alan Rew, eds. 1985. *Social Anthropology and Development Policy.* ASA Monograph No. 23. London: Tavistock Publications.

Hage, Per. 2004. East Papuan Kinship Systems: Bougainville. *Oceania* 75(2): 109–124.

Hoben, Alan. 1982. Anthropologists and Development. *Annual Review of Anthropology* 11: 349–375.

Jackson, Richard. 1991. Not Without Influence: Villagers, Mining Companies and Governments in Papua New Guinea. In *Mining and Indigenous Peoples in Australasia*, ed. J. Connell and R. Howitt. Sydney: Sydney University Press.

Jorgenson, Dan. 1997. Who and What is a Landowner? Mythology and Marking the Ground in a Papua New Guinea Mining Project. *Anthropological Forum* 7(4): 599–627.

Keesing, Roger. 1989. *Creating the Past: Custom and Identity in the Contemporary Pacific.* Honolulu: University of Hawaii Press.

Kirsch, Stuart. 2015. *Reverse Anthropology: Indigenous Analysis of Social and Environmental Relations in New Guinea.* Palo Alto, CA: Stanford University Press.

Laracy, Hugh. 1976. *Marists and Melanesians; A History of the Catholic Missions in the Solomon Islands.* Honolulu: University Press of Hawaii.

Lasslett, Kristian. 2009. *Winning Hearts and Mines: The Bougainville Crisis, 1988–90.* London: Routledge.

May, R.J., and M. Spriggs, eds. 1990. *The Bougainville Crisis.* Bathurst: Crawford House Press.

Nash, J. (1974). *Matriliny and Modernisation: The Nagovisi of South Bougainville.* New Guinea Research Bulletin No. 55, Port Moresby and Canberra, Australian National University, pp. 63–64.

Ogan, Eugene. 1971. Nasioi Land Tenure: An Extended Case Study. *Oceania* XLII(2): 81–93.

————. 1991. The Cultural Background to the Bougainville Crisis. *Journal de la Société des Océanistes* 92(1): 61–67.

————. 1996. Copra Came Before Copper: The Nasioi of Bougainville and Plantation Agriculture 1902–1964. *Pacific Studies* 19:31–52.

————. 1999. *The Bougainville Conflict: Perspectives from Nasioi*. Technical Report Discussion Paper, Australian National University, 99/3.

Oliver, Douglas. 1955. *A Solomon Island Society: The Siwai of Bougainville*. Cambridge: Harvard University Press.

————. 1991. *Black Islanders, A Personal Perspective of Bougainville 1937–1991*. Honolulu: University of Hawaii Press.

Palomka, Peter, ed. 1990. Bougainville: Perspectives on a Crisis. *Canberra Papers on Strategy and Defence*, vol. 66.

Simpson, S. Rowton. 1976. *Land, Law and Registration*. Cambridge: Cambridge University Press.

Skalnik, Peter. 1989. Lihir Society on the Eve of Mining Operations: A Long Term Project for Urgent Anthropological Research in Papua New Guinea. *Bulletin of the International Committee on Urgent Anthropological Research*, Nos. 32–33. Vienna: UNESCO.

West, Richard. 1972. *River of Tears: The Rise of the Rio Tinto Zinc Corporation Limited*. London: Earth Island.

Worsley, Peter. 1964. *The Trumpet Shall Sound*. London: Methuen.

# CHAPTER 3

# Panama and Indigenous Peoples

After the arrival of Rio Tinto (known at the time as RTZ or Rio Tinto Zinc) in Panama, an international campaign[1] was launched by the NGO Cultural Survival claiming that the company would be complicit in the disappearance of a unique indigenous culture with a 4000-year history.[2] The image of a large mining company destroying the cultural heritage of small groups of indigenous people was vigorously promoted by a number of NGOs. The campaign had public credibility in view of what had happened to Indians in Brazil and elsewhere in South America. Despite legislative protection, the treatment of Indian groups had been horrific, involving mistreatment, neglect, and threatened annihilation. NGO groups were against the development and wanted the Indians to be left alone.[3] The symbol of their cause was the *comarca*, which since 1972 had been a constitutionally protected area within which Indian land rights had to be respected. If the *comarca* was respected and not put to one side because of mining, then all would be well.

Any mining within the reservation area or *comarca*, was likely to be examined very closely by the NGO community.[4] The area was on the continental divide in the Province of Chiriquí, 260 miles west of Panama City. The copper deposit was located in mountainous terrain, steep slopes, and generally nutrient-poor soil with high rock content—characteristics that make farming difficult. On the Caribbean slope there was no dry season, and tropical forest dominated the landscape; on the Pacific slope there was

© The Author(s) 2017
G. Cochrane, *Anthropology in the Mining Industry*,
DOI 10.1007/978-3-319-50310-3_3

a windy dry season (December to April) and a wet season. Small perennial streams and larger rivers run on both sides of the continental divide and are used for bathing, laundry, and drinking. In that region most travel was done on foot or horseback, as there were few year-round access roads leading into the *comarca*. A mine access road ran up to Buàbti and continued to Escopeta, the location of the Cerro Colorado mine, from San Felix, a city connected to the Inter-Americana highway via Las Cruces.

Since 1932 it had been known that there were vast copper deposits in the Cerro Colorado area of Panama. The involvement of big international mining companies came about in the 1970s when the Panamanian leader General Torrijos, who had negotiated with US President Carter in 1977 to take control of the Panama Canal, decided to develop the Cerro Colorado copper deposit with the help of outside companies. Two of the early companies undertaking exploration were Texas Gulf and Canadian Javelin. Finally, the government entered into an agreement to develop the site with the mining company Rio Tinto Zinc of London. In 1979 RTZ began developing a proposal for a US$2 billion mine. Initially RTZ concentrated on trying to understand the ore body and the economics of mining in this part of the world. General Torrijos wanted the Cerro Colorado copper deposit developed in an exemplary manner and was particularly concerned to preserve the rights of the indigenous Guaymi People. He appointed his companion Lizia Lu, a University of California Berkeley–trained anthropologist, as the Director of the Social Development Department.[5]

The area was devoid of any physical infrastructure with the exception of that created for the exclusive use of the mine. The altitude of Cerro Colorado was between 650 and 1500 meters above sea level, and the ore body was near the center of the Isthmus, about 40 kilometers from both the Atlantic and Pacific Oceans. Politically, the Cerro Colorado project was located within an administrative subdivision of the state denominated *corregimiento*. This particular *corregimiento*, Hato Chami, incorporated within its boundaries 24 Guaymi communities of which 11 were situated within a 3-kilometer radius of the ore body.

As a first step toward mine development roads into the *comarca* mining camps were constructed, and serious geological testing commenced. But there was no formal agreement with the Guaymi that covered and gave permission for these new actions. The government's plans to develop the Cerro Colorado copper mine along the Cordillera Central in eastern Chiriquí Province gave impetus to the efforts of some Guaymi to organize politically. Guaymi attended a number of congresses to protect their claims to land and to publicize their misgivings about the projects. The Guaymi were concerned about the government's apparent lack of interest in their plight, about the impact on their lands and their productivity, and about the effect of dam construction on fishing and water supplies. Guaymi were also worried that proposed cash indemnification payments for lands or damages would be of little benefit to them in the long run.

In the tropical lowlands of Latin America indigenous peoples have resisted external encroachment on their traditional land areas. Mineral and other resource companies were moving into remote areas where, whatever the national law said about surface land rights, rights over subsurface resources have traditionally been vested in the state. Customary law may dictate patterns of land use and inheritance, but for minerals of strategic importance or high commercial value the state has reserved for itself the ultimate right to proceed with exploration and development. Many anthropologists believe that this is a danger because many indigenous communities do not appreciate the distinction between surface and subsurface rights.

But in the 1980s (and before the adoption of the UN Declaration on the Rights of Indigenous Peoples—UNDRIP—in 1997) this panorama was changing. In remote regions, indigenous organizations were becoming increasingly sophisticated, backed by influential international NGO and support groups. They were made aware of a current of emerging international law, which aimed to reinforce a concept of territorial rights based initially on the International Labor Organization (ILO, a specialist United Nations agency) definition. The ILO's Convention covers the total environment of the areas which indigenous and tribal peoples occupy or otherwise use. This philosophy explains the content of Article 15 of Convention No. 169, which aimed wherever possible to secure indigenous peoples' participation in the benefits of mining and other activities within their traditional lands.

## THE GUAYMI

The Guaymi living near the mine did not have a well-developed idea of how they should engage with the outside world. Living as they did in quite isolated settlements, spread out over large distances, lacking effective leadership, the Indians themselves lacked the institutions that would have enabled them to create their own view of the kind of development process in which they wanted to participate. There was no hereditary leadership;

the society was acephalous; such leadership positions as existed had been created for the purpose of easing communications with non-Guaymi. The *caciques* (chiefs) did not possess resources to distribute, which might tend to strengthen their position. The fragmented Guaymi economy and limited division of labor in traditional areas, which served to isolate and protect Guaymi culture in the past, also precluded the emergence of vigorous new leadership. With the assistance of missionaries and personnel from voluntary organizations, a number of meetings of the Guaymi nation were held, but these tended to reflect the agenda of the outsiders rather than the Indians.

A small number of traditional Guaymi were located in the High Sierra near the mine, the others living on the periphery of the larger society. The orientations of the two residential indigenous groups were different. This categorization was confirmed when, as the Indians saw things, outsiders made promises they did not keep, took pieces of their land, gave them what were considered to be the worst jobs, and made advances toward their women. A mine could increase the Indians' sense of inferiority quite dramatically. No matter what they seemed to try to do, the Guaymi found themselves in an inferior position. In housing, education, jobs, health, language, literacy, their position was inferior. The traditional Indian response was to withdraw. Withdrawal would no longer be possible with a mine in place.

In 1982 there were 88,000 indigenous people in Panama, of which Guaymi numbered 34,000. Twenty-five percent of the Guaymi lived within 20 kilometers of the mine site. They had an 85 percent illiteracy rate, 70 percent did not speak the nation's language (Spanish), 88 percent had not had any schooling, and 99 percent of those who did receive schooling had not completed primary school. Skin and respiratory diseases were common in the area, primarily because of lack of food and unhealthy sanitary conditions. Guaymi per capita income was around US$300 a year, or approximately 30 percent of the national average.

## Deteriorating Environment

Guaymi homelands within the *comarca* had been deteriorating year by year because of cultivational practices that had little to do with mining. In the 400 years since the Spanish had arrived, the Guaymi had been forced to move higher and higher into the mountains. As this relocation progressed, cattle and other ruminants were introduced. Feeding the cattle and horses encouraged the Guaymi to convert woodlands into pasture. Population increase further reduced the possibilities for traditional slash and burn exploitation of woodlands. The concomitant result produced changes to the lives of the Guaymi that had some similarities with the changes that accompanied the introduction of the horse to the Plains Indians in the USA.[6]

By the early 1970s it was becoming clear to those who studied the Guaymi and cared about them deeply that their way of life was changing and might not be preserved. Fieldwork by UN agencies such as the Food and Agricultural Organization and the World Bank showed that most of the problems of the indigenous Guaymi people had been caused by the uncontrolled environmental degradation of their homelands on the high continental divide between the Pacific and Atlantic oceans, and by the Guaymi coming into contact with strong pesticides when they worked on banana plantations in the lowlands.[7]

If left to themselves, would their lives get better? Large areas of the Guaymi homeland had undergone a process of desertification, the result of the growing inability of the hard laterite soils with no vegetational cover to retain water. Over 80 percent of the land was no longer suitable for cropping, and desertification was increasing. Excessive slash and burn exploitation of the environment was estimated, by the World Bank, to be capable of causing permanent damage to agriculture on the Chiriquí plains within a 20- to 30-year period. As soils on the steep slopes in the High Sierra became more and more unable to hold water, both flash flooding and drought would increase on the Chiriquí plains.

A diminished capacity for agricultural self-sufficiency resulted in an increasing dependency on the outside world for cash and goods. Guaymi society was neither fully engaged in slash and burn exploitation nor completely adjusted to sedentary agriculture. Part of the impetus for changing land-use patterns had undoubtedly come from rapid population increase and adverse land conditions. Guaymi agriculture had to try to adapt to the periodic absence of males, who worked on coffee farms and banana plantations. It was fairly obvious that the Guaymi were increasingly reluctant shifting cultivators. This had consequences for the amount of land they were able to exploit; that is, a smaller area had caused a speeding up of the rotational cycle.

Traditional Guaymi life and the opportunity to follow traditional patterns had long since vanished.[8] Exploitation of the changing environment called for new responses, and Guaymi lacked the skills and knowledge to realize that they were now harming their own environment. The most common crops grown were corn, rice, beans, otoe, bananas, and coffee, although people also grew tomatoes, peppers, and other vegetables in smaller gardens at home. Fruits such as mango, oranges, and nance grew seasonally along with cacao, all of which supplemented the Guyamí diet. Meat was rarely eaten, although many families kept cows, pigs, ducks, and chickens; sardines were a common staple.

The amount of food that could be grown was affected because of the need to adapt to new soil conditions; small and large ruminants introduced entirely new elements. New also was the need to have men travel long distances to earn money to supplement the living wrested from the barren soil at home. As the man-to-land ratios worsened, birth and infant death rates increased, resulting in the Guaymi, unlike indigenous groups located in the rain forests of the Amazon or the Philippines, becoming unable to continue to live as their forefathers had done.[9]

Housing illustrated what could happen when the Guaymi were left in splendid isolation. Traditional Guaymi housing made of stout timber and thatched with leaf had adequate flooring and possessed a high conical shape that took care of smoke from cooking fires. But from the time the Spanish arrived the Guaymi had been pushed higher and higher toward the continental divide. Ranchers took over from the Spanish and the Guaymi arrived at a point where their houses had to be sturdy since winds on the continental divide often reached 70–100 km/hr.[10] But traditional materials, wood, and leaf, were increasingly unavailable so Guaymi had to substitute sheets of galvanized metal. They had no model of a house built of galvanized metal, however, and so the results of their improvisation created more problems than were solved.

## EXIT

In the early 1980s RTZ decided to exit, and today the Cerro Colorado mine has still not been developed. The decision was taken on commercial grounds. When asked what might help the Guaymi living in the area I suggested that the most practical form of assistance would be an all-weather track suitable for four-wheel drive vehicles. RTZ did not have a dedicated "Community Team" to engage with the local population, but had relied instead on the government of Panama to look after community issues. Unfortunately, the pioneering work done by the Social Development team under Lizia Lu's leadership did not receive the recognition it deserved. The Social Development Department declined after the death of General Torrijos in a plane crash in 1981.

The Cultural Survival campaign focused on the wrong target. What was really needed was to spread information about the environmental situation and to try to help the Indian groups to adjust to new ways of earning a living. But the critics showed little enthusiasm for fieldwork and may not have had the hands-on experience and skills to identify and understand the full scope of the changes that were affecting the life of the Guaymi. When RTZ left and smaller companies became interested in developing the copper deposit, Cultural Survival's interest in an anti-mining campaign declined. Cultural Survival did become involved again when Guaymi working in the lowlands on banana plantations were said to be affected by the use of pesticides.

As Cerro Colorado illustrated, mining companies have often been attracted to areas where indigenous peoples live because their lands have had significant mineral deposits, as well as increasingly valuable water resources. NGOs and aid agencies have been concerned to protect these groups from exploitation and to do something about the worrying life statistics of indigenous peoples. Their mortality and morbidity patterns, their literacy rates, and their ability to make a living have all been significantly depressed in relation to mainstream society.[11] Their lives are shorter, and in many cases their birth rate is below that of the rest of the country where they live. The term "indigenous" covers groups large and small, wealthy and poor who all want choice rather than to be chosen; they may want some advantages of "civilized" technique, and some of the results of "civilized" knowledge, but over and over again, they have made it clear that they want to make use of improvements in a rhythm of their life in a community that they had inherited, even if it was a modified community.

## PROTECTION BY THE UN DECLARATION ON THE RIGHTS OF INDIGENOUS PEOPLES (UNDRIP)

The ILO was the first international body to address indigenous issues in a comprehensive manner by means of ILO 107, the Indigenous and Tribal Populations Convention in 1957 and ILO 169, the Indigenous and Tribal Peoples Convention of 1989.[12] ILO Convention 107 was phrased to encourage the integration of indigenous peoples into broader society, while 169 emphasized respect for the institutions, cultures, and rights of indigenous peoples and the right to determine their own development path. ILO 107 convention was ratified by 27 states and ILO 169 convention by 22 states.[13] Many of the countries that signed had no indigenous populations; others, such as the Latin American signatories, had an historic commitment to indigenous peoples, seeing them in Rousseau-like terms as noble savages, and gave little thought to what was needed to implement the conventions.

The UNDRIP that was passed by the UN General Assembly in 2007 was the culmination of many years of work by the indigenous peoples themselves at the UN in Geneva. It is ambitious, and it is as yet unclear whether resources will be available to indigenous peoples for implementation. The UNDRIP put indigenous peoples in the driving seat when it came to who made decisions about their future. Prior to the Declaration, indigenous communities were seen as being unable to handle their own affairs. After UNDRIP, the idea expressed in Article 22 of the Covenant of the League of Nations thinking that "Native Peoples who were unable to stand on their own under the strenuous conditions of the modern world" needed "safeguards" became redundant.[14]

Australia, New Zealand, the USA, and Canada (Canada signed up to UNDRIP in 2010) voted against adoption of the Declaration as did African countries that did not believe in the whole ideology of indigeneity. The Declaration is not a legally binding document. States will, however, be expected to pass supporting national legislation, as was the case with the 1948 Declaration on Human Rights. In North America the US Supreme Court has ruled that certain Indian groups already have the legal status of sovereign nations. New Zealand has the treaty of Waitangi, while Australian and Canadian courts have recognized indigenous land rights. Yet in a number of South American countries the constitutional safeguards for indigenous peoples have performed poorly.

How is indigeneity defined? Self-definition is usually taken to be the most important element in deciding whether a people is or is not entitled to be called indigenous. No single definition exists capable of capturing all indigenous peoples. The United Nations used the definition of indigenous people applied by José Martinez-Cobo, the Special Rapporteur to the Sub-commission on Prevention of Discrimination Against Indigenous Populations. Cobo states that "Indigenous communities, peoples and nations are those which having a historical continuity with pre-invasion and pre-colonial societies that developed on their territories, consider themselves distinct from other sectors of societies now prevailing in those territories, or parts of them. They form at present non-dominant sectors of society and are determined to preserve, develop and transmit to future generations their ancestral territories, and their ethnic identity, as the basis for their continued existence as peoples, in accordance with their own cultural patterns, social institutions, and legal systems."[15]

Following the Report of the Human Rights Council in 2006, General Assembly voting was delayed as a result of an initiative by Namibia. African countries had little sympathy for the indigenous peoples' movement and threatened to vote against or abstain en masse. After a year of intense lobbying, African nations secured amendments that balance what the indigenous peoples have demanded with the power of the state. The Declaration went further than the ILOs' Convention 169 which was ratified by only a few countries, most of which had no indigenous peoples. ILO 169 left the ownership of subsurface mineral rights to the state. The UN Declaration says the mineral rights—and a new claim, water rights—belong to indigenous peoples. Article 11 goes beyond physical forms of heritage to include cultural, religious, intellectual, and spiritual property. Legacy issues will be difficult to sort out when mines have been in operation for many years. In such instances, indigenous peoples have suggested that companies should acknowledge past problems and develop common goals for the future.[16] The Declaration contained provisions for communities to be involved in the approval of development proposals, to have the right of return related to resettlement, and for communities to manage their own cultural heritage programs.

What resources are available for enforcement? The legal route to implementation of the Declaration, while helpful, was unlikely to be enough. Some years ago the government of the Philippines brought out excellent rules covering Ancestral Domains which included a requirement to consult with indigenous peoples. But since no administrative resources were allocated to assist with implementation, the measure did not meet its purpose.[17] Contemporary thinking has no provision for courts or tribunals or other deliberative bodies, or even government, for that matter.

It is not enough to say that such and such a government has ratified because, in the case of ILO 169, for example, substantial ratification was claimed, but many of those countries had no indigenous populations and, beyond signifying consent, their approval does little for implementation. Implementation has to have local roots, and any judgments need to be located much closer to the people concerned than in a UN Committee. Unfortunately, in the attempt to gain visibility and authority, the institutions that are relied upon are globally rather than locally visible, and to date, little has been done to involve local governments.

Those who want to assist in achieving broader recognition of what is needed for UNDRIP need to secure changes in the way the problems of indigenous peoples are viewed in a number of countries. For example, in Sarawak, as we shall see later, public opinion held that the indigenous Penan should be made to embrace development for their own good so that, as one politician told me, "they should not give birth in the jungle." Until and unless that public perception of an indigenous life that, while culturally attractive, should give way to modern ways of living is addressed and changed, it is unlikely that the spirit and the intent of well-meaning international action by the ILO and other agencies will make a real difference.

In societies where the gap between the way of life of broader society and the indigenous peoples is perceived to be very wide, as with the Penan or the Sri Lankan Veddah, progress will be slow until a more enlightened view takes hold, and that is not likely if the argument is framed in highly emotional terms such as the "bad exploiters" and "victims of the miracle" that have often characterized the debate in South America.[18] India has useful lessons because the situation there with respect to the public understanding and sympathy for tribes and scheduled castes differs in several respects from the situation encountered in other parts of the world. In Canada or Australia, the European settlers are of relatively recent provenance, whereas in India, coexistence has a history of several hundred years. Whereas plunder and physical mistreatment were the order of the day elsewhere, in India the British took it upon themselves to protect and preserve the Andaman Islanders, the Nagas, the Todas, and others. At the time of independence in 1947, the framers of the Indian Constitution continued this tradition of taking special measures to protect tribal peoples. Some 7 percent of all union budgets in India is set aside for tribal people.

Looking at a map of India today one can see that the amount of land set aside for tribal peoples is very large indeed. The Indian tradition of protecting tribal people was also continued after independence in 1947. Decisions affecting tribal peoples were still left to officers of the Indian Administrative Service. It is still said to be the case today that what the collector says commands greater public respect than the decisions of governors or other politicians. Recently, the 1997 Samantha judgment, given by three Indian Supreme Court Justices, ruled that the alienation of tribal lands to mining companies, or even the government, was illegal. Tata Steel, one of India's most respected companies, would be affected because its mines and plant are on tribal lands. Under the Fifth Schedule to the Indian Constitution, however, tribal lands could be alienated by a state, and since Samantha, developers have been trying to use this loophole.

Indigenous peoples are those best placed to come up with solutions to their own problems. In Canada, nobody but the First Nations was able to halt the terrible inroads caused by alcoholism in areas where monies from extractive industry obviously did more harm than good to their recipients. It was the Indians themselves who started the Healing Lodges and it is the Indians who have led the pushback against deterioration. Indigenous peoples have to help themselves and have to work out their own means of salvation.

Experience suggests that indigenous peoples have three choices in their search for an accommodation with the outside world, none of which can provide them with any certain insulation against the effects of social change. They can try to resist all change while attempting to perpetu-ate selected elements of their culture. Examples of this approach would include "nativistic movements," such as the Ghost Dance of the Sioux. But this, although peaceful, caused panic among surrounding settlers and led to the slaughter of indigenous people without much cause. Attempts by indigenous people to address wrongs done to them led to the Battle of Wounded Knee in the United States and the Maori Wars in New Zealand which ended with the Treaty of Waikato in 1840.[19] More recently, in 2011, bows and arrows were taken up by tribal peoples in Orissa in their protest against a project by the Norwegian company Norsk Hydro. Police fired on the crowd of protesters and there were deaths. Large sums of money had been paid but without help to the recipients to prevent gam-bling and alcohol from consuming funds. Apparently all the funds were paid without considering the negative impact that this could have on local prices for basic household items. At Keonjihar, a town in Orissa, an Australian missionary and his children were burned to death in their car in 1999 by Hindu fundamentalists who had accused the missionaries of trying to proselytize. Orissa remains a state where there is much suspicion of Christian missionaries.[20]

Such movements usually fail to halt change. Outside influences, mis-sionaries, health care, education, are usually too strong to resist. A recent example of this kind of fairly futile thinking is the stated intention of the International Union for the Conservation of Nature and Natural Resources to establish "protected anthropological areas." These would be "areas set aside for the continuance of ways of life endangered by the expansion of industrial civilization and its technology. They are areas occupied by people practicing ways of life of anthropological or historical importance and are intended to provide for the continuance of these ways of life for so long as there are people willing to practice them and capable of doing so."[21] But who is to judge that continuance is not possible? This is an important question because many indigenous groups are fragmented and bitterly divided by the effects of alcohol and family feuds and the pressures created as a result of their shrinking resource base.

The second option would be to embrace radical change, a move accompanied by conspicuous abandonment of well-known cultural symbols. This prepares the present and future generations so that they can achieve a smooth transition or assimilation into mainstream society. It is a strategy reminiscent of China's "Great Leap Forward" or Attaturk's changes to Ottoman society. The third option, and probably the most common, would be to try to appreciate and preserve the local cultural heritage while acquiring the knowledge and skills required to journey to the outside world in order to earn a living for varying periods. North American Indian groups have been doing this with some success for many years. One example is that of the Mohawk, a deeply divided community who have the international border between Canada and the USA running through their reservation. They follow "Long House" traditions but their elected councils are not recognized by large segments of the community. Nevertheless the Mohawk are doing better than other First Nations who are operating casinos. They have retained a fierce pride in their cultural heritage and they have developed a talent for working in the construction business, erecting steel girders at great heights in large cities. Of course, neither indigenous society nor nearby society can be expected to stand still; there will inevitably be contact with the outside world that produces change. Such change process cannot be stopped. What is essential is that the trajectory of change is monitored.

On the issue of engagement with the outside world, Marcia Langton, who has done a great deal to recognize what is right and what has been wrong in the way mining companies have dealt with Aboriginal people in Australia, provided a case that exemplifies the tensions faced by a young man working for a mining company. She asked a young Aboriginal man working on a mine site if he enjoyed his job. He said it was good to be out of the community for a couple of weeks and away from the fighting.[22]

Mining companies are discovering that some groups want to be recognized as indigenous people because of the protective advantages such a status provides. Recently the Metis, a people of mixed French and

Indian parentage, made representations to the Canadian government that would enable them to achieve indigenous status. In Namibia the Rehoboth Basters, a group of African, Afrikaaner, and German parentage, prepared an appeal to the UN to be recognized as an indigenous people because they thought that this would offer them superior protection in their ongoing struggle with the Namibian authorities. But the concept of indigeneity has not been welcomed in Africa, the western states of the USA, or in PNG.

## INDIGENOUS LAND RIGHTS UNDER THE LAW OF THE STATE

When European settlement of Australia began in 1788, it was convenient for the British colonists to assume that the territory they entered was not owned by anyone. This view was enshrined in the idea of terra nullius. Governor Bourke interpreted it in 1835 to mean that indigenous Australians could not sell or assign land, nor could an individual person or group acquire it, other than through distribution by the Crown. Joseph Trutch, the first Lieutenant Governor of British Columbia, insisted that First Nations had never owned land, and thus could safely be ignored. That was an overly optimistic assessment since the First Nations in the area had rich natural food sources such as salmon, dense populations with permanent villages, and complex systems of rights and statuses. There are many other examples of terra nullius claims. Svalbard was considered to be a terra nullius until Norway was given sovereignty over the islands in the Svalbard Treaty of 9 February 1920. The Philippines and the Peoples' Republic of China both claim the Scarborough Shoal or Panatag Shoal or Huangyan Island, nearest to the island of Luzon, located in the South China Sea. The Philippines claims it under the principles of terra nullius and EEZ (Exclusive Economic Zone) while China's claim refers to its discovery in the thirteenth century by Chinese fishermen.

In 1982 five Torres Strait Islanders, including Eddie (Koiki) Mabo, began legal proceedings to establish title for the Meriam people to Murray (Mer) Island. The historic 1992 landmark Mabo decision in the High Court of Australia rejected the doctrine of terra nullius, the notion being that Australia did not belong to anyone at the time of European settlement. This rejection of the concept meant the Australian legal system recognized that indigenous ownership of land may have continued after the British colonization of Australia.[23]

When the High Court of Australia ruled that Australian law recognizes "native title," it meant that the law recognizes the ongoing existence of these customs and traditions through which indigenous people have a connection to their land. "Native title" is not a new form of title created by the High Court but legal recognition of those rights that Aboriginal and Torres Strait Islander people have always had. The High Court also confirmed that when Britain gained sovereignty over Australia, it gained "radical title," but not "full beneficial title," to the land. This means that while the government has the ultimate power over the land, it does not automatically gain full ownership of it. The government is able, however, to extinguish native title to land when it uses its sovereign power in a way that shows a "clear and plain intention" to do so. In the Mabo decision, the High Court of Australia said that native title had been extinguished on all freehold land and certainly the vast majority of leasehold land.

In 1996 the High Court of Australia made another important decision in the Wik case which relates to a claim of native title on land that included pastoral leases granted by the Queensland government. The High Court of Australia said that native title can only be extinguished by a law or an act of the government which shows clear and plain intention to extinguish native title. The laws creating pastoral leases in Queensland did not reveal an intention to extinguish title. The High Court of Australia found that Queensland pastoral leases had been created to meet the needs of the emerging pastoral industry. The rights and interests of a pastoral lease-holder had to be determined by looking at the relevant statute and at the lease itself. This process showed that the leases in question did not give the leaseholders a right to exclusive possession of the land. Therefore, the granting of a pastoral lease did not necessarily extinguish native title. Native title could exist along with the rights of the leaseholder.

The Supreme Court of Canada's decision in the Delgamuukw case in late 1997 confirmed that aboriginal title did exist in British Columbia, which entails a right to the land itself—not just the right to hunt, fish, or gather. The Gitxsan Nation and the Wet'suwet'en Nation initiated the lawsuit in 1984. The court held that First Nations did have title to their lands and that they could engage in non-traditional activities so long as these did not damage their cultural heritage. Aboriginal title applied to communal territories because individuals could not hold or sell land. Tribal law would prevail, and the courts would listen to oral histories and songs in any legal proceeding, to determine ownership. Despite this landmark decision, it was clear that negotiations, consultation, and litigation might go on for a number of years.

The High Court of Canada's decision concludes with these words: "Ultimately, it is through negotiated settlements, with good faith and give and take on both sides, reinforced by judgments of this Court, that we will achieve ... the reconciliation of the pre-existence of aboriginal societies with the sovereignty of the Crown."[24]

In Australia:

Country, to use the philosopher's term, is a nourishing terrain. Country is a place that gives and receives life. Not just imagined or represented, it is lived in and lived with. Country in Aboriginal English is not only a common noun but also a proper noun. People talk about country in the same way that they would talk about a person: they speak to country, sing to country, visit country, worry about country, feel sorry for country, and long for country. People say that country knows, hears, smells, takes notice, takes care, is sorry or happy. Country is not a generalized or undifferentiated type of place, [...] rather, country is a living entity with a yesterday, today and tomorrow, with a consciousness, and a will toward life. Because of this richness, country is home, and peace; nourishment for body, mind, and spirit; heart's ease.[25]

## TRADITIONAL LAND TENURE

Ownership of land held under traditional tenure is best seen as a bundle of rights.[26] There are two aspects. The first is the nature of tenure, and tenure is a matter of what is held, the rights that are held, and by whom they are held. Individuals may have rights of use, but it is usually the family, the clan, or the lineage that holds the right to use or to bar use. The second aspect is how these rights are held. In most traditional societies this is a matter of following normal customs and practices on the part of the individual, family, clan, or other unit. Striking differences may exist.[27] One of the distinctive features of land held under customary title by indigenous peoples is the fact that it is "owned" by all members of the indigenous group or subgroups thereof rather than single individuals. Land registration is a process involving a survey of the physical dimensions of the land and then recording them in a Deeds Register.[28] Systems of land registration do not provide for registering the names of all members of a lineage, clan, or other group. For administrative and practical convenience, they require ground to be registered in the names of a few males.[29] But then there is the practical question of which males should be chosen and how? Whereas in the legal systems of industrialized countries inheritance takes place on the death of an owner, in traditional societies inheritance may often take place at birth.

How does one choose, as PNG legislation in 1962 suggested, six males out of 100 or more, even when the system is matrilineal and should therefore acknowledge women's names and their brothers' rights? The extinguishing of the rights of those in whose names ground is not registered can be a time bomb, one that is best diffused by recognizing the problem at the outset and trying, as was done in PNG, to register the ground in the name of the lineage or group with traditional ownership rights.

It is extremely difficult to transfer "ownership" of ground held by a group under traditional title to an individual under a system of primogeniture. In order to comply with traditional law where "ownership" is vested in a lineage, clan, or other social group that may have 100 or more members, any transaction affecting traditional grounds would have to have been understood and agreed by all members of the social group, including very young children. If one woman were pregnant, those living could not ignore the interest of the unborn child. More difficult still is the fact that the dead may also on occasion be considered to have ownership rights. After all, the dead are usually held to be the most senior generation in many societies and, as such, they are usually thought to be against change.[30]

Whereas the thinking dominant in industrialized countries suggests that an individual can sell or dispose of his or her interest in land, in traditional societies the individual—who is like a shareholder—has no identifiable estate to sell or lease. When trying to look at traditional tenure through our eyes, the most obvious fact is how limited and circumscribed the rights of individuals are with respect to what they can do with land. The attitudes and beliefs just mentioned obviously make it difficult to turn land into a commercial commodity, especially where traditional beliefs are still strong. The response has been to ignore traditional rights and to convert land held under traditional tenure into ground held under our principle of primogeniture thereby ensuring that the ground can be bought and sold and used as collateral for loans. But vast areas of land have not been surveyed and registered in poor countries.[31]

In spite of the passage of UNDRIP and landmark cases in Australia and Canada there has been no indication that the government in either country would launch a large mapping initiative to record land boundaries and traditional law. In many countries, including Australia and Canada, an investigation and documentation of each aboriginal society's land tenure rights may only take place when triggered by a developer's wish to use ground, and then that assessment will usually be concerned to ensure that land acquisition or use would be legal from the point of view of the dominant legal system. Yet there are many First Nations and aboriginal groups whose land boundaries and traditional law have not been established and recorded. Any mapping and recording exercise would obviously be a contested, and resource-intensive, process. As time passes, there will be fewer and fewer First Nations and aboriginal people left alive who are knowledgeable about their traditional legal system.

A complication to the establishment of a better understanding of traditional law in Australia, for example, might come from the fact that the number of Aboriginal people living in towns is increasing, which may mean that more and more Aboriginal people are no longer connected to the land. Since it is assumed that a traditional way of life is needed to ensure the effective functioning of traditional law, what has been the effect of urbanization on traditional legal norms? The members of an Aboriginal group judged to be land-entitled may never have operated on a face-to-face basis in day-to-day community living. In other parts of the world land acquisitions have been challenged many years after their execution, as the Indian claims in New York State in the USA have demonstrated.[32]

Indigenous legal systems do not necessarily share Western ideas about the physical characteristics of land. Unlike the situation in Western societies indigenous legal systems may not see property rights extend to the stratosphere or the center of the earth, rivers may not have a notional divide in the middle of a watercourse, and low and high tide may have no meaning. Who owns what in the river can also be of great importance if and when panning is introduced into rivers carrying tailings. Perhaps strangest of all, trees and ground may have different owners. Considerable misunderstanding and occasionally loss of life has been caused by locals and surveyors not realizing that they had different ways of looking at land.[33] Misunderstanding has also been created between geographers and anthropologists over where boundaries should be located.[34] In both Africa and Asia, difficulties have been created as a result of local people having a very different conception of the properties of land from that of the company. Instead of thinking of land as being like a box with the boundaries being considered the four sides, local people conceived of their ground as being defined by its center. The center was frequently associated with a shrine. Cultivation took place around the central point. When surveyors and their employers staked out boundaries according to their perception of land, having four sides, like a box, this caused problems with local people.[35]

In some communities, when land was bought it was sometimes later found that economic trees such a coconut, cocoa, or coffee bushes or even trees that produced good firewood belonged to someone other than the owner of the ground. Surface water rights need special attention and it may be even more important to understand underground ownership and use.

## ACQUIRING LAND FROM TRADITIONAL GROUPS[36]

The conflict over land and the criticism of indigenous groups over the past 40 or 50 years suggests that the ways in which land acquisitions have been conducted by miners has contributed to the negative feeling that has been created against the industry, both locally and internationally. Consequently, in many instances it may make sense to try to ensure that any land transactions are not only legal from the point of view of a state's laws but also legitimated in terms of the customary and traditional legal norms of the community that is involved.[37]

Was the purchase of Manhattan Island for a few beads legal from a traditional perspective? Almost certainly not. The deeply emotional nature of land tenure for indigenous peoples is indicated by the fact that in a large number of Pacific island societies the word for "land" is "placenta." In the Solomons I had to take a man by ship from Sinerangu to the administrative center at Auki on the west of Malaita island; he had just been speared as a result of a land dispute that had begun 50 years earlier. He died shortly after reaching hospital.

Relationships that demonstrate respect and concern for customary ways are appreciated in traditional societies where men and women not only have to converse in two or more languages, but may also have to take account of two legal systems, the dominant belonging to the state and the local belonging to the community. It is essential that the local legal system is not ignored just because of the power of the state. Mining companies have had a bad reputation with indigenous peoples particularly in the USA where, for example, the Western Apache had reservation lands taken from them and given to mining companies, some of which refused to hire Apache workers.[38]

For miners, understanding the nature of social change and how this affects traditional tenure is of critical importance. In Australia after the Mabo case the law was intended to facilitate interaction between Aboriginal people and outsiders wishing to lease or use land held under traditional title. Yet the concepts that were used, such as "The Traditional Owners," may not always have encouraged mining companies to look more closely at traditional law.[39] It might be helpful to know the extent to which traditional law and marriage customs are being observed; the extent to which sharing customs still work in urban households; and the effectiveness of positive and negative sanctions on providing the sorts of social control that communities want.

In Australia the concern with gaining access to the land of Aboriginal people has been to satisfy the land acquisition requirements of the state rather than traditional concepts of land tenure and use. When miners make what they hope are long-term leasing and benefit arrangements with Aboriginal people the result will be legal from the point of view of Crown law, but will these agreements continue to be seen as consistent with traditional law and fair by Aboriginal people? Following UNDRIP, which gives backing to those who feel they may be able to claim that they were coerced into deals that were not legal in traditional terms, it makes sense for mining companies to do a risk analysis of their land acquisitions. The extent to which traditional law has been understood and applied to these state-regulated and approved acquisitions is not clear. How will "land connected peoples" regard the loss of that connection in 15 or 20 years' time?

## Muslim Tenure

When indigenous land is held by individuals as in Muslim countries then very different issues must be considered to those encountered with group tenure in indigenous societies. When ground is acquired from communities where Muslim tenure prevails in Serbia or the Balkans, for example, special care has to be taken to understand what may happen to those whose entire land holdings are reduced or lost because of landlessness. The fragmentation of holdings, particularly in the Muslim countries, has been an increasing concern of governments and developers for many years. Increasing family size has played a major role in this fragmentation process which eventually sees families dependent on earning their living from very small parcels of land that are no longer economic to work. In Sri Lanka, for example, a Deeds Register had an entry for a three-hundredth of a Yak tree. However, experience suggests that land law is seldom applied in a mechanical manner and that there is usually some wiggle room provided the way custom, tradition, and land law are combined in actual inheritance cases. It is in this space that mining companies need to see what rich and poor people do locally and on the basis of this information they must fashion solutions which preserve the family and the position of women while identifying ways of sustaining livelihoods.[40]

The egalitarian traditions of Islam informed the early systems of land tenure in Muslim countries with tenets of social justice and equality, and in the nineteenth century the Ottoman administration codified the prevailing land tenure systems into a set of regulations that are still followed in much of the Middle East. Yet it is also the case that customary tenure continues to regulate a large part of the access to, and the use of, the land; but these customary land tenure systems are failing to address age-old problems: landless households and small farmers continue to compete for limited and fragmented cropland, and pastoralists are losing control of their traditional grazing areas, and meanwhile the gap between the big landowners and the poorest elements of society continues to grow. Agricultural economists argue that the greatest productivity gains come when producers own the land they work, but many depend on communal lands and here is also the problem of irrigation. Access to water is becoming an increasingly important issue as the number of users grows. Land and water tenure cannot be separated. In desert and semi-desert areas, pastoral production needs a carefully structured system of drinking water sources for both humans and animals.[41] In the case of crop production, water access needs to be regulated in line with its scarcity.[42]

## LAND TENURE IN PERU

La Granja, a copper deposit located in the Cajamarca region in Northern Peru, was first discovered in the late 1960s. Rio Tinto was the third mining company to own the right to evaluate the La Granja deposit via a Transfer Agreement with the government of Peru. Cambior, a Canadian-based mining company no longer in existence, held the La Granja concession from 1993 to 1999. During their tenure, Cambior sought to gain land ownership and implemented a land acquisition and relocation program. According to community members, the company used pressure tactics (with the support of the Peruvian national government) to close down schools and health clinics in order to convince landowners to sell their land in exchange for cash compensation.

About half the population chose to sell their land and leave. The majority of those who left moved to the Lambayeque Region on the coast, and many of the relocated families faced significant challenges in reestablishing their livelihoods in the new settings. The experience with Cambior created conflict and trauma within and between families and a shared social memory of what they felt was an unfair land acquisition process. In 2002, BHP Billiton sold the 2600 hectares back to the displaced families at below market prices, using the entirety of the proceeds to fund a foundation designed to facilitate the reconstruction of the communities. Families were given the option of relocating back to the area to rebuild their homes and villages.

Rio Tinto had to face an uphill battle because with the lease from the Peruvian government came legacy issues. Long before Rio Tinto arrived, the area had been plagued by violence and civil unrest. Miners such as Rio Tinto were regarded with considerable suspicion by local people. The trauma associated with the Cambior land purchase from 1995 to 1997 weighed heavily in the collective experience of the communities. La Granja's future was greatly dependent on obtaining community consent to land access for drilling as well as community consent for voluntary land acquisition.

The most important thing before any offer was put on the table was to understand local land tenure and to engage in the slow and patient building of local relationships characterized by trust. Over time community consensus has been skillfully created and used to encourage constructive engagement with key individuals. The investment risks were considerable: the resettlement team that was formed in 2010 was selected not only because of their own expertise and knowledge of resettlement but also because of the ability to form and maintain relationships with the local people throughout the project area. With an experienced Community Team on the ground, problems and risks could be anticipated and nipped in the bud. The technical and social teams shared the same office, helping to maintain dialog and facilitating the collaborative approach. This may appear to be a simplistic point, but it is more often the case in mining to have the technical design teams based in another company, another office, or even another country. Physical distance between the social and technical staff can hinder internal engagement, limiting the understanding of the situation on the ground and the identification of risks to the development of a particular design.[43]

## NOTES

1. See Brysk (1996).
2. David Maybury-Lewis, founder of Cultural Survival, whom I had known for some time told me when he heard I was going to look at the Guaymi for RTZ that I would need "a long spoon." And see Gjording (1981).
3. See Dostal (1972).
4. The *comarca* or reservation inside which Indians were to be protected was enshrined in the Panamanian Constitution, and this is also the case in a number of other South American countries. The idea, Rousseau-like in its conception, was that ensuring splendid isolation for indigenous peoples such as the Guaymi of Panama was the way to preserve a unique culture. However, the states with this provision in place did little to enforce it and the intent of the legislation was frequently breached.
5. See Greene (1984).
6. See Hamalainen (2008).
7. See Utting (1993). The environmental situation is not reflected in, see Gjording (1981) or Young (1971).
8. See Bort (1981).
9. Figures based on statistics compiled by Food and Agricultural Organization (FAO), United Nations Children's Fund (UNICEF), Ministries of Health, Planning, and Agriculture, as well as the Social Department of CODEMIN.
10. See Jaen (1982).
11. See Hall and Patrinos (2013).
12. In the USA indigenous groups had Treaty rights with the US government which gave them the status of sovereign nations in a number of respects. Tribal sovereignty refers to tribes' right to govern themselves, define their own membership, manage tribal property, and regulate tribal business and domestic relations; it further recognizes the existence of a government-to-government relationship between such tribes and the federal government.
13. See Henriksen (2008).
14. See Callahan (1999).
15. See Cobo (1987).
16. See Coates (2013), Toensing (2013).
17. See Tuminez (2013).
18. See Davis (1977).
19. See Linton (1943).
20. On Naxalite fundamentalists see their activities in Nepal in Lawoti and Kumar (2009).
21. See International Union for the Conservation of Nature (n.d.).
22. Marcia Langton, Third Boyer Lecture, Melbourne, December 2012. I am indebted to Paul Wand and Bruce Harvey for letting me see the innovative approaches that they have brought to Aboriginal employment in Rio Tinto operations in Australia.

23. See Russell (2005).
24. Discussed in Salazar and Alper (2001).
25. See Bird (1996).
26. See Cochrane (1974).
27. See the chapter on law in Cochrane (1971a).
28. See Dowson and Sheppard (1956).
29. See Simpson (1976).
30. See Cochrane (1971b).
31. See, with respect to turning land into collateral, see de Soto (2000).
32. See Campisi (1976).
33. In the Solomons on the island of Malaita surveyors were attacked by local people and one missionary was speared to death. However, in Africa where the same focal concept exists it does not seem to have caused the same same problems. For example, the plateau Tonga have had a focal concept of land. See Colson (1958).
34. See Sutton (1995).
35. See Cochrane (1969).
36. See Bhuta (1998).
37. Very useful information can be obtained from the Land Tenure Center at the University of Wisconsin, Madison, Wisconsin.
38. See Perry (1996).
39. See Holcomb (2004).
40. See Forni (2002).
41. Some sense of the complexities is provided by Wittfogel (1956).
42. See Ruttan (1986), Wade (1987).
43. This land acquisition was skillfully managed by Sharon Flynn. Louis Cononelos, David Salisbury and the author served as external advisors to Sharon in Peru. See Flynn and Vergara (2015).

**Acknowledgments** Professor Richard Perry provided advice and comment related to indigenous issues in North America and Roger Plant who drafted ILO 169 provided information on South America.

## BIBLIOGRAPHY

Bhuta, N. 1998. Mabo, Wik and the Art of Paradigm Management. *Melbourne University Law Review* 22(1): 24–41.

Bird, Deborah Rose. 1996. Land Rights and Deep Colonising: The Erasure of Women. *Aboriginal Law Bulletin* 3(85): 6–13.

Bort, J.S. 1981. *An Environmental Reconnaissance of the Cerro Colorado Concession Area. Empresa de Cobre Cerro Colorado*. Panama City: Societe Anonyme.

Brysk, Alison. 1996. The Internationalization of Indian Rights. In *Latin American Perspectives 2(3)*. Ethnicity and Class in: Latin America (Spring).

Callahan, Michael D. 1999. *Mandates and Empire: The League of Nations and Africa, 1914–1931*. Brighton: Sussex Academic Press.

Campisi, Jack. 1976. The New York-Oneida Treaty of 1795: A Finding of Fact. *American Indian Law Review* 71–82.

Coates, Ken, ed. 2013, September 18. Ken Coates, and Terry Mitchell, eds., *From Aspiration to Inspiration: UNDRIP Finding Deep Traction in Indigenous Communities*. The Center for International Governance Innovation (CIGI).

Cobo, José Martinez. 1987. *Study of the Problem of Discrimination Against Indigenous Populations*. UN Document E/CN.4/Sub.2/1986/7.

Cochrane, Glynn. 1969. Choice of Residence in the Solomons and a Focal Land Model. *Journal of the Polynesian Society* 78(3): 330–343.

———. 1971a. *Development Anthropology*. New York: Oxford University Press.

———. 1971b. Juristic Persons, Group and Individual Land Tenure: A Rejoinder to Goodenough. *American Anthropologist* 73(5): 1152–1155.

———. 1974. Land Alienation: The Case for Traditionalists. *Oceania* XLV(2): 124–131.

Colson, E. 1958. *Marriage and Family Among the Plateau Tonga of Northern Rhodesia*. Manchester: Manchester University Press.

Culhane, Dara. 1998. *The Pleasure of the Crown: Anthropology, Law, and First Nations*. Burnaby, BC: Talon Books.

Davis, Shelton. 1977. *Victims of the Miracle: Development and the Indians of Brazil*. Cambridge: Cambridge University Press.

de Soto, Hernando. 2000. *The Mystery of Capital, Why Capital Triumphs in the West and Fails Everywhere Else*. London: Bantam Press.

Dostal, W., eds. 1972. *The Situation of the Indian in South America*. Geneva: World Council of Churches.

Dowson, E., and V.L.O. Sheppard. 1956. *Land Registration*. 2nd ed. London: HMSO.

Flynn, Sharon, and Liz Vergara. 2015. *Land Access and Resettlement Planning at La Granja*. CSRM Occasional Papers: Mining Induced Resettlement Series, Centre for Social Responsibility in Mining, Queensland University, St. Lucia, Brisbane.

Forni, N. 2002. *Land Tenure Policies in the Near East*. Rome: FAO.

Gjording, C. 1981. *The Cerro Colorado Copper Project and the Guaymi Indians of Panama*. Occasional Papers No. 3. Cambridge, MA: Cultural Survival.

Greene, Graham. 1984. *Getting to Know the General*. London: Bodley Head.

Hall, Gillette, and Harry Anthony Patrinos. 2013. *Indigenous Peoples, Poverty and Human Development in Latin America*. New York: Palgrave Macmillan.

Hamalainen, Pekka. 2008. *The Comanche Empire*. New Haven: Yale University Press.

Henriksen, John B. 2008. *Key Principles in Implementing ILO Convention No. 169*. Geneva: ILO.

Holcomb, Sarah. 2004. Traditional Owners and 'Community Country' *Anangu*: Distinctions and Dilemmas. *Australian Aboriginal Studies* 2: 64–71.

International Union for the Conservation of Nature. n.d. *World Directory of National Parks.* International Union for the Conservation of Nature: Glans, Switzerland.

Jaen, Bernardo. 1982. *El Impacto Del Project De Cerro Colorado in El Pueblo Guaymi y Su Futuro.* Centro De Estudio Y Accion Social: Panama.

Lawoti, Mahendra, and Anup Kumar. 2009. *The Maoist Insurgency in Nepal: Revolution in the Twenty-first Century.* London: Routledge.

Linton, Ralph. 1943. Nativistic Movements. *American Anthropologist* 45(2): 230–240.

Perry, Richard. 1996. *From Time Immemorial: Indigenous Peoples and State Systems.* Austin: University of Texas Press.

Russell, Peter. 2005. *Recognizing Aboriginal Title: The Mabo Case and Indigenous Resistance to English-Settler Colonialism.* Toronto: University of Toronto Press.

Ruttan, Vernon W. 1986. Assistance to Expand Agricultural Production. *World Development* 4(1): 39–63.

Salazar, Debra J., and Donald K. Alper, eds. 2001. *Forging Truces in the War in the Woods.* Seattle: University of Washington Press.

Simpson, S. Rowton. 1976. *Land Law and Registration.* Cambridge: Cambridge University Press.

Sutton, Peter. 1995. *Country: Aboriginal Boundaries and Land Ownership in Australia.* Aboriginal History Monograph No. 3. Canberra: Australian National University.

Toensing, Gale. 2013. Political Party! Celebrating UNDRIP and Indigenous Culture in Montreal. *Indian Country Today.*

Tuminez, Astrid. 2013. *This Land Is Our Land: Moro Ancestral Domain and Its Implications for Peace and Development in the Southern Philippines.* Baltimore, MD: Johns Hopkins, School for Advanced International Studies.

Utting, Peter. 1993. *Trees, People and Power: Social Dimensions of Deforestation and Forest Protection in Central America.* London: Earthscan.

Wade, Robert. 1987. The Management of Common Property Resources: Finding a Cooperative Solution. *The World Bank Research Observer* 2(2): 129–133.

Wittfogel, Karl August. 1956. *The Hydraulic Societies.* Chicago: Chicago University Press.

Young, Philip D. 1971. *Ngawbe: Tradition and Change Among the Western Guyami of Panama.* Chicago: Board of Trustees of the University of Illinois.

# CHAPTER 4

# Miners Join the UN Global Compact

In 2000 the UN Secretary General Kofi Annan launched a massive private sector initiative when he invited the CEOs of the world's largest corporations to join a Global Compact whose main object was to prevent corruption, avoid human rights abuses, and protect the environment while implementing the Millennium Development Goals (MDGs).[1] There was some irony in the fact that in arranging for the CEOs to police themselves the UN was following a strategy known as Indirect Rule that had been developed by British Colonialism in West Africa. There Lord Lugard came up with the idea of "Indirect Rule," which involved making the traditional chiefs part of the apparatus of administration and control. This enabled the British to rule vast areas and large populations without requiring large staff numbers.[2] Now it looked as if the UN intended to manage mining company CEOs in the same way.[3]

In 2000 the United Nations Development Programme (UNDP) announced Global 2000, an initiative with big business participation aimed to bring a billion poor people to the market place. NGOs challenged the inclusion of business in the Global Compact. NGO leaders wrote to UNDP to complain that this initiative would allow some of the world's nastiest companies to wrap themselves in the UN flag. UNDP abandoned the program, after suggesting that they had intended to reform the sinners and surely it was reasonable to assume that only by working with these transgressors could there be hope of reform.[4]

The enthusiasm with which NGOs approached the issue of surveillance of the mining industry as a 24/7 issue brought to mind an idea of the eighteenth-century English political thinker Jeremy Bentham who

© The Author(s) 2017
G. Cochrane, *Anthropology in the Mining Industry*,
DOI 10.1007/978-3-319-50310-3_4

was interested in penal reform. In 1787 he put forward the idea of the Panopticon. This supposed that the relationship between the guards who needed to watch the prisoners and the inmates themselves could be made much more efficient if, in the center of the facility, a tall tower (the Panopticon) were to be constructed. The watchers' location at the top could not be seen from the ground. However, sufficient light would be available, day or night, to permit the surveillance of every cell and thus it would create the ability to know what every inmate was doing at all times. Inmates would never know if one or more guards were watching them and as a result they would learn to behave as if they were under continuous surveillance. As Michel Foucault said about Bentham's idea, he who is subjected to a field of visibility, and who knows it, assumes responsibility for the constraints of power; he makes them play spontaneously upon himself; he inscribes in himself the power relation in which he simultaneously plays both roles; he becomes the principle of his own subjection.[5]

As the protest over Global 2000 showed, NGOs wanted business performance to be under continuous NGO surveillance because the mining industry was not to be trusted. The Global Compact incorporated a transparency and accountability policy known as the Communication on Progress (COP). The annual posting of a COP was an important demonstration of a participant's commitment to the UN Global Compact and its principles. Participating companies were required to follow this policy, as a commitment to transparency and disclosure was thought critical to the success of the initiative.[6] Once big business had signed on to the Global Compact it was always going to be difficult to determine how much should be done and what would be done. Although business was not anxious to assume responsibility for all the problems that governments would rather not tackle, how could limits be set? Traditionally, businesses have been reluctant to become involved with UN agencies and public sector organizations that they have found to be slow-moving and inefficient. But, given the trust deficit between business and the public, companies knew they had to comply.

The UN Global Compact was the largest corporate sustainability initiative in the world, with 20,000 signatories based in more than 135 countries. The UN Global Compact was a call to companies everywhere to (1) voluntarily align their operations and strategies with 10 universally accepted principles in the areas of human rights, labor, environment, and anti-corruption, and (2) take actions in support of UN goals, including the MDGs. The UN saw mining as a business that made money in a way that harmed communities. Whatever extractive industry was doing in the community it was obviously wrong and needed to be changed. A growing volume of complicity charges had involved the miners in Bougainville and Ok Tedi, the oil and gas industry in Alaska and the Gulf of Mexico making it clear that extractive industry needed watching.[7]

## A Lack of Emphasis on Relationships

Surveillance was not all the UN was interested in. Key policymakers in the UN system believed that extractive industry could do more to help with the implementation of the UN agenda for developing countries. By the turn of the century UN development agencies and NGO believed that they needed to harness business resources because it was becoming clear that Third World governments were not delivering the results that were expected. One did not have to look too hard to find a senior International Agency official or an NGO leader who believed that Third World governments were failing to deliver much-needed growth and services and that improvement was highly unlikely.[8] Concern about the performance of Third World government was joined to concern about the performance of extractive industry.

Economists in the UN believed that the success of MDGs and Sustainable Development Goals (SDGs) depended on a development process whose results could be quantified and shown to improve individual well-being. A Columbia University paper suggesting that the SDGs should be seen as being close to the core business objectives of mining companies. The UN's great improvement initiatives, the MDGs and the SDGs, focused on improving the lives of the poorest individuals on the planet. Unfortunately, neither the MDGs nor the SDGs did much to accommodate relationships. As one observer says[9]:

> One of the major concerns expressed from a number of quarters about the SDGs (SDGs have now replaced the MDGs)[10] regards the underlying worldview which focuses so much on individual rights. This view grows out of Western legal decisions and practice rather than reflecting the balance of responsibilities and obligations recognized in much of the so-called "third world." Equally, there is concern about the language of "developing countries" and "developed countries" which carries an implicit message that the determinant of a society's level of progress is measured primarily by its level of wealth and income rather than by the characteristics of its broader culture. So the SDGs, as currently articulated, fit uneasily with the traditions and values characteristic of many of those in Africa, Asia and Latin America.[11]

With the MDGs and SDGs emphasis was placed on the use of individual indicators of development similar to those associated with individual well-being in the industrialized countries. The Victorians, imagining they represented the apogee of civilization when comparing themselves to savages, chose as indicators of progress those attributes such as cleanliness and Godliness, which characterized their own supposed fitness to rule. Income and health tick-box data do measure dimensions of poverty, but they have the additional, and unfortunate, effect of suggesting that the problems of the poorest and their solutions are known. The result of this sort of analysis is that solutions then become a matter of supply. Why not global supply?

For some, poverty is like AIDS; it produces a state of increasing social disconnectedness, that is, an absence of relationships, for the afflicted individual. The MDGs and SDGs looked at poverty in terms of their low incomes and poor health and other aspects which concentrated on what it was that poor people did not have. The community relations approach asked about the assets poor people had, such as strength, dexterity, and personal qualities that could contribute to success. Nobody can escape from poverty unless they can do something for themselves. Even a soup kitchen cannot be useful if people cannot walk to it. Of course, the poor need medicines, money, education, and food, but the material deprivation, degradation, and squalor of their existence is not all that those wishing to help need to appreciate. It is necessary to know the extent to which groups and communities in dire need can help themselves. Knowing what they can bring to any improvement attempt is essential if assistance is to succeed. Viewed from afar, the poor do not seem to have a culture or heritage to be understood, protected, or preserved. They do not seem to have intellectual property worthy of study, any admirable physical characteristics, or jokes anyone might to repeat. They appear to have too little of everything except children.

The poor can be very far away indeed from help because many of the things that poor people do not have, besides purchasing power, are taken completely for granted by the rest of society. We need to imagine a quite different world. Miners were aware that the more severe the poverty, the greater the degree of social disconnectedness for the individual involved. Social atrophy, which accompanies poverty, can affect speech, food preparation, personal hygiene, and other kinds of behavior whose form and content are derived from social interaction. Poverty occurs in society, but it is often not of society. As with HIV/AIDS, its growth cuts the individual off from society. It has many forms, some controllable, some not, some completely debilitating, some not. Helpers need to begin with this single person and the need to understand his or her situation in order to try to reattach this individual to a family, community, or society by means of a relationship.

## HUMAN RIGHTS

A UN human rights initiative was driven by the NGO focus on big business not small, low-profile, state-owned companies.[12] The UN initiative expected companies to work on human rights where much of the implementation responsibility had been previously executed by UN agencies and the International Labor Organization (ILO). Child labor and workers' rights provided an example of this problem. These human rights had been on the books for many years without effective enforcement and without the ILO assuming much responsibility for implementation.

Prior to involving mining companies in its human rights agenda the UN had not calibrated what free speech or cruel and unusual punishment meant at community level in developing countries and business was not in a position to start from scratch to do such a job. From a community perspective the absence of any appraisal of what are locally considered to be human rights at the community level was and still is a massive impediment to a useful implementation process. Comparative data collection of the sort represented by the Human Relations Area Files (HRAF) program at Yale was lacking, and no effort was made to expand coverage. Alison Renteln Dundes illustrated problems that can be caused when universal human rights are applied in traditional communities and when what communities consider to be human rights violations are overlooked.[13] Anthropologists have debated whether or not there can be universal rights because of the belief that cultures are unique and cannot be compared.[14] Substantial disagreement began with the 1947 Statement on Human Rights of the American Anthropological Association. Arguments for and against have been recently revisited by Clifford Geertz.[15]

There was no requirement from the UN or NGOs to ensure that communities had participation and voice in this exercise. There was no community survey which could have illustrated the number and seriousness of human rights offenses, which universal rights had local currency, and which deeply felt local human rights needed to be recognized. The human rights initiative failed to accommodate situations where wrongdoing was attributed to groups and collective entities rather than individuals. In India idols had legal personality. In PNG witchcraft was seen as being far more important than most of the crimes contained in the Universal Declaration. Tradition in Kiribati allowed men to bite off their wives' noses if they suspected infidelity. In the Solomon Islands local people were incensed when the government made adultery a civil offense to be pursued by the aggrieved party, whereas if adultery were seen as a criminal offense the government would do the prosecuting since criminal acts are those that, in theory, upset everyone (which was the case with adultery). Trade union representatives worried that a collective approach to human rights approach could help companies to weaken collective bargaining by expanding individual work place contracts.[16]

In a number of developing countries there was a hierarchical legal system with the top level responsible for the 1947 Universal Declaration while at the lowest level native courts served traditional complaints.[17] How in the absence of due diligence were companies supposed to understand and deal with local processes and personal law procedures such as Adat in Indonesia[18] or conflict between Hindu and Muslim law in India?[19]

The growing number of alleged human rights abuses on the part of business had earned a strong UN response. In 2011 and as part of the Global Compact group of initiatives, Professor John Ruggie, the UN's Special Representative on Human Rights, laid out what companies were expected to do. He said that the state should protect and that companies should respect human rights.[20] What was not made particularly clear to those in the mining industry who were presented with Ruggie's work was the extent to which, in the UN family, there were overlapping responsibilities, jurisdictions, and contradictions, and there was an overall absence of tried and tested methods of implementation. There was no supporting body of case law.

In the hands of mining industry critics human rights proved to be a very elastic concept. Under pressure from NGOs the UN was rapidly expanding its definition of human rights abuses but not doing enough to ensure that there was effective implementation. The state had the responsibility to provide access to remedy through judicial, administrative, and legislative means, and corporations had the responsibility to prevent and remediate any infringement of rights to which they contributed. Having effective grievance mechanisms in place was thought to be crucial in upholding the state's duty to protect and the corporate responsibility to respect. The guiding principles say that non-judicial mechanisms, whether state-based or independent, should be legitimate, accessible, predictable, rights-compatible, equitable, and transparent. At the operational level companies were encouraged to operate through dialog and engagement, rather than with the company acting as the adjudicator of its own actions.

The ICMM queried the Ruggie proposals, by saying, "It is not accepted that a survey of 65 NGO-reported instances of alleged abuses provides any balanced or fact base for differentiating the performance of different sectors. Campaigning NGO often select issues on which to focus based more on whether the sectors or companies involved are publicly visible or have high-profile brands, rather than on any comparative analysis of corporate human rights performance. In this respect, any overall judgment regarding the extractive sector needs much better substantiation."

Business had supported the Voluntary Principles on Security and Human Rights scheme promoted by the British Foreign Office that focused on the use of private security personnel. Business was not anxious to become a vehicle for UN global diplomacy by taking over what was clearly in many instances a government responsibility. There was a lack of clarity over the boundaries between companies and states in upholding human rights (while seeking to uphold human rights within their legitimate "sphere of influence," for example, companies needed to avoid becoming political actors, or looking as if they were interfering in the political affairs of host countries).

Ruggie presented his "Protect, Respect, Remedy" framework to the UN Human Rights Council in 2009 with very strong support from NGO activists.[21] Although it was not perhaps appreciated at the time, the "Protect, Respect, Remedy" framework Ruggie had given to the miners was very similar to the "Respect, Protect, and Fulfill" provisions of the 1966 Covenant of Economic, Social and Cultural Rights (CESR) where it was insisted that states' parties which had ratified the Covenant had accepted an obligation to implement it.[22] The CESR's language which said that states had an obligation to "fulfill" their human rights obligations seemed to become Ruggie's suggestion that companies should "remedy" human rights wrongdoing. If true this represented the transfer of matters that had been a responsibility of the public sector onto the private sector. Inevitably, there were questions about the competence of business to fulfill this role and how the private sector would be able to avoid appearing to be prosecutor, judge, and jury in matters where its own performance was in dispute.[23]

While it was obviously useful, right and proper to upgrade the approach of mining companies and their employees to their human rights duties under the law the attempt to make these companies a global vehicle for the implementation of that law and the responsibilities of UN agencies and states was bound to end in a certain amount of disappointment among those with a genuine interest in helping the UN. What was not made as clear as it should have been was the fact that all sorts of labor issues that were formerly the remit of the ILO became human rights issues that were passed to business. A very heavy implementation burden was placed on mining company personnel working in the community. Almost none of the communities' team were lawyers and none had received any form of legal training.

Thereafter, mining companies spent a great deal of time logging and discussing industrial relations disputes, sexual offenses, and assaults, many of which were criminal in nature and had little to do with human rights. The implementation burden has grown as industry critics have begun to label every small wrongdoing a human rights violation because that categorization gets much more attention. As time passed the human rights tick-box enforcement burden began to consume up to half of the time of community personnel; this meant that other tasks could not be addressed. Very few company personnel had contacts with those enforcing human rights in the countries where they were operating.

It might have been helpful if the implementation of the Ruggie proposals had followed Section 44 of the Indian Constitution which said that new human rights should only be introduced on the basis of local demand. The necessary due diligence was not done. Ruggie did not say, nor did NGOs and UN agencies say, that (they and) companies should begin by trying to understand the local legal scene.

Because there was limited understanding among the NGOs and UN agencies of the legal situation in remote communities there was, and still is, confusion over what companies and governments should do and how they intend to work with business. Ruggie's work created an uncertainty which has not been effectively countered over the nature and extent of the rights that business is expected to uphold. Grievance mechanisms were established for communities even though it was clear that they might throw up issues which were the responsibility of government and no coordination mechanisms had been established. Some companies began to construct a parallel system to that of the state while others did not realize that much of what they were doing was not the responsibility of business but of the state.

## GLOBAL REPORTING

Businesses in the Global Compact were required to report annually on their progress with the implementation of the initiative's 10 principles covering human rights, workplace standards, the environment, and anti-corruption. Consecutive failures to submit a COP leads to removal (delisting) of the company and some 2000 companies have now been delisted. "Over the years, the Global Compact's framework has become increasingly robust, through the establishment of integrity measures, introduction of guidance materials, and the support of the many local networks that provide COP mentoring," said Georg Kell, Executive Director of the UN Global Compact. "By all indications, we expect corporate disclosure of policies and practices to become more common, as companies joining the Global Compact increasingly enter with a better understanding of the critical value of reporting on environmental, social, and governance performance."[24]

To implement global surveillance the United Nations Environment Programme (UNEP), UNDP, and NGOs began sending out questionnaires. To do this they began by developing questionnaires that encouraged box-ticking against global standards for ethical corporate behavior which they themselves had invented, usually without reference to those in the communities who were to be surveyed. They reduced, simplified, and standardized complex and highly variable community situations overseas in order to have quick and effective communication with their fundraisers and the general public. All businesses—from dry cleaning and hotels to mining—all over the world were asked to respond to the same questions. No matter that the businesses had vastly different footprints and the countries had vastly different ideas of proper business behavior.

The surveyors assume that what needs to be known about various communities overseas can be done by questionnaire or even a short survey of a few days, weeks, or even a couple of months. It is a belief that, by identifying common factors in a problem, one can arrive at a common solution. If people do not have access to clean potable water, water must be provided; if people have a low income, more money must be provided; if people are illiterate they need education; if people are malnourished better food will be provided; if people are ill-served by their government, better governance will be provided. However, because individuals and communities differ millions of short surveys would be needed to begin to understand a little about poverty in the world. Then millions of re-surveys would be needed to keep such surveys current. Moreover, does it make sense to have one set of people—and usually the wrong people at that—saying something superficial about what the poor need if there is nobody in place to do much about the problems identified?

The UN Global Compact resulted in Rio Tinto and other mining companies having to agree to participate in a system of global surveillance covering the environment as well as human rights and community relations. Over the next decade the international reporting system was put in place by big NGOs and UN agencies for the Global Compact and other initiatives to make sure that the miners behaved in a socially responsible manner. The tick-box questionnaire responses were reminiscent of Armchair Anthropology. Over 100 years ago anthropologists wrote books about "savages" on the basis of armchair reflection. Instead of visiting the savages, the authors designed questionnaires in their studies and sent them to reliable travelers and others in the field in the hope that, once completed, these would confirm what the armchair anthropologists had thought in the first place.[25] Since it was commonly supposed that savages were promiscuous, any salacious details about savages tended to be believed and repeated without any of the armchair anthropologists bothering to test the quality of the information. Today, tales that conform to the popular stereotype of the wicked global corporation tend to be believed in the same way. None of the eminent armchair anthropologists thought that it was actually necessary to go and live with savages in order to understand them. When asked by a visitor if he had ever met any of the savages he had described so convincingly, Sir James Frazier, the early Cambridge anthropologist replied, "Good Heavens, no!"

Two events were instrumental in persuading these armchair anthropologists to move out of their comfortable surroundings. One was the expedition by Cambridge University to the Torres Straits in 1898; the other was the publication of Bronisław Malinowski's work during World War I in the Trobriand Islands in what is now part of PNG. The development of good community skills and understanding, which anthropologists call fieldwork, gradually became recognized as requiring them to live in a community long enough to learn the language and customs of those people they were trying to understand.

A further level of surveillance was provided by the UNEP Global Reporting Initiative (GRI) which big business was pressured to join. GRI believes that reporting on economic, environmental, and community performance by business organizations is as routine and comparable as financial reporting. GRI was established in 1997 and is managed by the UN Environmental Agency.[26] It is a successor organization to the Coalition for Environmentally Responsible Economies (CERES) and represents a coalition of over 50 investor, environmental, religious, labor, and social justice groups. Questionnaires are structured around a CEO statement; key environmental, social, and economic indicators; a profile of the reporting entity; descriptions of relevant policies and management systems; stakeholder relationships; management performance; operational performance;

and a sustainability overview. The data is compiled by the participating organizations, who must involve stakeholders in the process. GRI now has over 1000 participating businesses but no aid agencies or NGOs.

GRI's "social" category includes the organizations' policies and performance with respect to human rights, labor practices, occupational health and safety, training and education, community performance, and product safety. The questionnaires seek to understand if there are standard policies in place and, if there are, whether there have been achievements or criticisms. GRI is much less concerned with positive contributions that might be made by a business than it is with corruption and anti-competitive behavior. The GRI questionnaire has over 100 indicators related to various categories of wrongdoing that might be committed by a company.

Extractive industry is now being invited to accept the surveillance of yet another reporting regime: the Initiative for Responsible Mining Assurance (IRMA). As was the case with GRI, companies will feel pressure to participate in yet another tick-box exercise. The IRMA will, like many other watching organizations, develop standards and offer participating companies certification. Founded in 2006 by a coalition of NGOs, businesses purchasing minerals and metals for resale in other products, affected communities, mining companies, and trade unions, the IRMA is developing standards for environmental and social issues related to mining, including labor rights, human rights, indigenous peoples and cultural heritage, conflict response, pollution control, and site closure. The IRMA expects to begin certifying mine sites in 2016 so that companies can support the mining of minerals and metals that is consistent with healthy communities and environments, and that leaves positive long-term legacies.

## NGO Campaigns

Campaigns provided miners with a vivid illustration of the power of NGOs to get what they wanted even in instances where they got the science wrong. A classic example was provided by the 1995 Brent Spar incident. This involved a dispute between the Shell Oil Company and Greenpeace over the disposal in the North Sea of an abandoned 14,500-ton oilrig platform. Shell had commissioned over 30 studies showing that disposal at sea was the best option. In addition it was clear that disposal on land posed a greater risk of injury as a result of working at height to dismantle the rig. But the studies that had been done were not publicized and Greenpeace, which orchestrated the campaign against Shell, had never been sufficiently close to the platform to be able to survey its contents and assess the risks.

The British government gave Shell permission for disposal at sea. Greenpeace then mounted a campaign to persuade European governments and the public that sea disposal posed an unacceptable environmental hazard. The Greenpeace campaign was based on strong visual images which had prime-time viewing on evening TV in countries bordering the North Sea.[27] What attracted viewers to the Greenpeace position was the idea that their marine environment—which most viewers imagined themselves visiting at some time in the future—was in danger of being polluted. Europeans sitting at home were also influenced by the Greenpeace campaign images of activists climbing onto the abandoned platform to save pristine ocean from unthinking harm and damage while fire hoses were used to deter them. Similar imagery has won the day for Greenpeace when protesting against whaling.[28]

Media reports portrayed the activists as having a deep concern for the environment and public welfare, and this was in sharp contrast to the coverage of Shell's company executives, who were portrayed as people who wanted to save money by dumping the platform at sea and degrading the marine environment.[29] The activists talked up the marine threat from disposal at sea that they said was posed by chemicals on the rig as well as a great deal of oil which could seep into the sea and damage the marine environment. This statement was later shown to be wrong—there were no chemicals and very little oil.[30]

Shell was subjected to commercial pressure; gas stations were boycotted and politicians, including Angela Merkel, who was Germany's Minister for the Environment, demanded that Britain cancel the permit it had given Shell for sea disposal. In the end Shell caved in and, much to the disgust of the British government, agreed to dispose of the platform on land in Norway. The rig ended up in a small community which was not consulted before it was decided that it would provide a final home for it. The platform was broken up and used as part of a new jetty in the harbor.[31]

As time went by the big NGOs did less global campaigning and small NGOs mounted vigorous campaigns against specific projects in specific countries. However, it was seldom the case that these protests by the small NGOs could be transformed and turned into global campaigns.

## A Suspicious Relationship

Where mining companies were concerned NGOs shared with Max Weber a deep suspicion as to the motives of those who pursued private profit, and an assumption that local people would, inevitably, be the victims of profiteering, environmental damage, and cultural vandalism.[32] Consequently, NGOs believed that they had the right to intervene in any mining operation anywhere in the world because they were "interested and affected parties." This version of carpet-bagging was thought all right in spite of the NGO not knowing the community, speaking their language, or having been asked by the community to assist.

Around the world small local NGOs attacked mining companies but not always without cost. In a number of African countries governments which regulated and licensed mining did not always take kindly to NGOs making comments suggesting that the wrong decision had been made by the authorities when they granted licenses and permissions. Of course, there was still a small number of NGOs who would never be convinced that there was any good in mining. One, called PARTIZANS, People against Rio Tinto Zinc And All Its Works, attended an annual general meeting of shareholders after the company name had been changed from RTZ to Rio Tinto plc to complain that it should be compensated for the expense caused by the name change!

In recent years mining companies have tried harder to develop their relationships with NGOs, and they have had some success with partnerships related to flora and fauna and small birds. However, community-level partnerships have proved to be more difficult to establish. And these difficulties suggest that if mining companies wanted a useful NGO partner it would be best if they started their own. For example, in 2000 the Kelian Equatorial Mining Company (KEM) in Kelimantan had to confront the fact that thousands of its neighbors were starving as a result of forest fires which had destroyed their crops. They had no food, and much of what remained was being eaten by rats. Out-of-control forest fires had killed the snakes that had kept the rat population under control. Wanting to distribute food and willing to pay the costs, the company wanted to work with an NGO on distributing food. But KEM found that to do this it had to start its own NGO to handle the food distribution because neither the big international, nor the local, NGOs would or could do the job. One large London-based NGO explained that in some countries they were prepared to cooperate with mining companies but in others they would not cross the street to meet with them.

There is no doubt that the future welfare of many remote communities will continue to depend on NGOs of various kinds continuing to play key roles in their development processes.[33] Many of the positive changes in development policy and practice that have taken place in recent years can be associated with the increased role and profile of NGOs. Although NGOs have become established organizational actors within development policy and practice, they share with miners the fact that critical questions are increasingly being asked of their performance and accountability.

## NOTES

1. See Williams (2004).
2. See Perham (1937).
3. That is not the only vestige of colonialism that miners can see overseas. Local NGOs frequently complain that the big international NGOs have a colonial attitude toward local NGOs and regard them very much as if they were the junior partner.
4. UNDP's opportunism was rebuffed in a way that made mining companies somewhat suspicious of that agency.
5. See Semple (1983), Foucault (1995).
6. See Kuiper (2004).
7. Critics of extractive industry—which includes mining—have pointed to major losses of life and catastrophic damage to local society. See Annex and Kirsch (2006), Pakinkas et al. (1993). See Stockman et al. (2005), Varma (2005), Swartz (2010).
8. See Rotberg (2003). See also, Migdal (1988).
9. See Schluter (2012).
10. The Millennium Development Goals Report by the UN in 2015 reported uneven progress and disappointingly limited impact on extreme poverty. There are 17 goals and 169 social indicators in the UNs new Sustainable Development Goals.
11. See Schluter (2016).
12. ICMM Submission to the UN March, 2006 on Ruggie's proposals.
13. See Dundes (1990, 2004), Tobin (2014). Despite the title Tobin does not do well dealing with the problem of applying Human Rights to traditional communities.
14. See Wilson (1997).
15. See Geertz (1984).
16. See Politakis (2007), Kolben (2010).
17. See the legal hierarchy described by Harvey (1966).
18. See Hooker (1978).
19. See Derrett (1978).
20. See The Kenan Institute for Ethics (2012).
21. See United Nations (2011).
22. See Alston and Quinn (1987).
23. See Rights and Accountability in Development (RAID) (2015).

24. See Hoessle (2014).
25. Armchair Anthropology is covered in Stocking (1996).
26. See Esty and Ivanova (2001). Paper makes the point that UNEP lacks legal authority and resources and as a result has not been able to hire high-quality manpower.
27. Campaigns involve short intensive bursts of intensive effort and depend on being able to harness what Ronald Knox called "enthusiasm," an energetic vision created by society that is more important than self. See Knox (1950).
28. Campaigns have been successful when they have appealed to "a sense of place." Attempts to tap underground water in areas with limited rainfall and imprecise understanding of the size of the underground resource or its replenishment rate have been sharply challenged by campaigns talking about taking "fossil water."
29. David Korten said, "We might reflect on what might have been accomplished over the past three decades if there had been more development movements and fewer development projects. It would be interesting to know more of the dynamics of this process. True movements draw their energy and resources from the people and have little definable organizational structure." See Korten (1986).
30. See Melchett (1995).
31. See Lofstedt and Renn (1997).
32. Max Weber, quoted in Mayer (1955).
33. See Moore (1998).

## BIBLIOGRAPHY

Alston, Philip, and Gerard Quinn. 1987. The Nature and Scope of States Parties' Obligations Under the International Covenant on Economic, Social and Cultural Rights. *Human Rights Quarterly* 9(2): 156–229.

Annex, A., and Stuart Kirsch. 2006. *Reverse Anthropology: Indigenous Analysis of Social and Environmental Relations in New Guinea*, 272. Palo Alto, CA: Stanford University Press.

Derrett, J.D.M. 1978. *Essays in Classical and Modern Hindu Law*. Leiden: Brill.

Dundes, Alison Renteln. 1990. *International Human Rights: Universalism Versus Relativism*. London: Sage.

———. 2004. *The Cultural Defense*. New York: Oxford University Press.

Esty, Daniel, and Maria Ivanova. 2001. Making International Environmental Efforts Work: The Case for a Global Environmental Organization. Paper prepared for an open meeting of the Global Environmental Change Research Community, Rio de Janeiro.

Foucault, Michele. 1995. *Discipline and Punish: The Birth of the Prison*. New York: Vintage Books.

Geertz, Clifford. 1984. Anti Anti-Relativism. *American Anthropologist* 86(2): 263–278.

Harvey, W.B. 1966. *Law and Social Change in Ghana*. Princeton, NJ: Princeton University Press.

Hoessle, Ulrike. 2014. The Contribution of the UN Global Compact towards the Compliance of International Regimes: A Comparative Study of Businesses from the USA, Mozambique, United Arab Emirates and Germany. *Journal of Corporate Citizenship* 53: 27–60.

Hooker, M.B. 1978. *Adat Law in Modern Indonesia*. Oxford: Oxford University Press.

Knox, Ronald A. 1950. *Enthusiasm*. New York: Oxford University Press.

Kolben, K. 2010. Labor Rights as Human Rights. *Virginia Journal of International Law* 50: 1–31.

Korten, David C. 1986. Introduction. In *Go To The People*, ed. James B. Mayfield, xi. West Hartford, CT: Kumarian Press.

Kuiper, Andrew. 2004. Harnessing Corporate Power: Lessons from the UN Global Compact. *Bulletin on the Development of Federalism* 47: 9–19.

Lofstedt, Ragnar E., and Ortwin Renn. 1997. The Brent Spar Controversy: An Example of Risk Communication Gone Wrong. *Risk Analysis* 17(2): 131–135.

Mayer, J.P. 1955. *Max Weber and German Politics*, 131. London: Faber and Faber.

Melchett, P. 1995. Green for Danger. *New Scientist* 148(2010): 50–51.

Migdal, Joel S. 1988. *Strong Societies and Weak States: State-Society Relations and State Capabilities in the Third World*. Princeton, NJ: Princeton University Press.

Moore, M. 1998. Corporate Governance for NGO? *Development in Practice* 8(3): 335–342.

Pakinkas, L.A., et al. 1993. Community Patterns of Psychiatric Disorders After the Exxon Valdez Oil Spill. *American Journal of Psychiatry* 150(10): 474–478.

Perham, Margery. 1937. *Native Administration in Nigeria*. Oxford: Oxford University Press.

Politakis, George P. 2007. *Protecting Labor Rights as Human Rights: Present and Future of International Supervision*. Geneva: International Labor Organization.

Rights and Accountability in Development (RAID). 2015. *Rethinking the UN Guiding Principles and Company Grievance Mechanisms*. Oxford: RAID.

Rotberg, Robert I. 2003. *When States Fail: Causes and Consequences*. Princeton, NJ: Princeton University Press.

Schluter, Michael. 2012. What Charter for Humanity? Defining the Destination of Development. In *After Capitalism: Rethinking Economic Relationships*, ed. Paul Mills and Michael Schluter, 62–73. Cambridge: Jubilee Centre.

———. 2016. Three Relational Concerns about the Sustainable Development Goals. Sustainable Development Goals: The Missing Dimension, A Relational Thinking Dialogue, Geneva, April. Unpublished Discussion Paper.

Semple, Janet. 1983. *Bentham's Prison: A Study of the Panopticon Penitentiary.* Oxford: The Clarendon Press.

Stocking, George W. Jr. 1996. *After Tylor: British Social Anthropology 1888–1951,* 15–24. London: Athlone Press.

Stockman, Lorne, James Marriott, and Andrew Rowell. 2005. *The Next Gulf: London, Washington and Oil Conflict in Nigeria.* Washington, DC: Constable & Robinson.

Swartz, Spencer. 2010. BP Provides Lessons Learned From Gulf Spill. *The Wall Street Journal,* September 5.

The Kenan Institute for Ethics. 2012. *The U.N. Guiding Principles on Business and Human Rights: Analysis and Implementation.* Durham, North Carolina: Kenan Institute for Ethics, Duke University.

Tobin, Brendan. 2014. *Indigenous Peoples, Customary Law and Human Rights— Why Living Law Matters.* London: Routledge.

United Nations. 2011. *Implementing the United Nations "Protect, Respect and Remedy" Framework.* New York and Geneva: United Nations.

Varma, Roli. 2005. The Bhopal Disaster of 1984. *Bulletin of Science, Technology and Society* 25(1): 37–45.

Williams, Oliver F. 2004. The UN Global Compact: The Challenge and the Promise. *Business Ethics Quarterly* 14(4): 755–774.

Wilson, Richard. 1997. *Human Rights and Cultural Context: Anthropological Perspectives.* London: Pluto Press.

# PART II

# Community Relations

CHAPTER 5

# Headquarters Inception

Rio Tinto was engaging with communities in the hot tropics, frozen tundra and the arctic, deserts, and high rainfall areas of the world. The neighbors of a mine might be white-collar executives in the USA, herders in Iran, subsistence agricultural tribesmen in New Guinea, snowmobilers and trappers in the Yukon, sheep farmers in New Zealand, Aboriginal hunters in Australia, or Dyaks foraging in the tropical woodlands of Borneo. In the industrialized countries miners may only want to be seen as a good neighbor, nothing more; in developing countries however, where advances in living standards are more problematic and government services of recent provenance, there are always pressures for a company to become involved in a wide range of betterment initiatives, some of which have at most a tenuous connection with mining.

Why would a mining company try to have good community relations? A senior Rio Tinto manager known for his hard-headed pragmatism said, "The maintenance of good relationships with communities is as important as the maintenance of the plant and the ore body." Good community relationships could avoid or reduce expensive delays and help the company to move more quickly through the regulatory hurdles; new investment opportunities went to companies that had a good track record in the community. Being seen to take community issues seriously had a positive impact on a company's ability to attract and retain talent because the best and the brightest wanted to work for companies that took social issues seriously. Shareholder value was boosted because investors liked companies with a good reputation.

© The Author(s) 2017                                                      83
G. Cochrane, *Anthropology in the Mining Industry*,
DOI 10.1007/978-3-319-50310-3_5

The intellectual challenge for Rio Tinto was to develop a community engagement process which was capable of universal application while accommodating cultural diversity. A first step was to consult with those in Rio Tinto who had experience of relationships with communities to try to understand if there were common lessons and recommendations. The opinions and advice of academics, critics, and community members were added to the mix. These consultations helped to develop an understanding across the company about how community relations had worked, what professionals saw and thought about the issues that were of concern, and the best way forward. From a community perspective what was wanted was not about having either global skills or local skills; it was instead about achieving a sensible balance in the interests of doing more to represent community perspectives and priorities in ways that provided local choice and creativity.

There was no shortage of opinion among those at corporate headquarters about what should be done in the community to improve community performance. Surely it was all simple? What could not be handled by a well-worded memorandum from London? Operations had so many Standard Operating Procedures (SOPs) at mines, why not another? An SOP is a set of step-by-step instructions created by a business to help workers carry out routine operations.[1] Their purpose is to achieve efficiency, quality output, and uniformity of performance, in the belief that the repeated application of unchanged processes and procedures and its documentation will produce the best outcome. Providing sites with simple tool kits was attractive to some corporate managers. But the tool kit idea while increasingly popular in the mining industry would do too little to improve skills, and its use has always seemed to me to be like hoping that cars could be fixed in a garage whose owner had not bothered to hire and train any mechanics.[2]

Reliance on the comforting linear process of the engineer discounted events and personalities, got over the fact that the executive might know little about the community where the operation was located, downgraded the chance of things not going according to plan, and banished the unacceptable thought that the executives involved should admit that they did not know what they were doing and that there was no shame in asking for advice in order to come up with something that would be more realistic. When applied to communities the assumptions were often heroic: all inputs became outputs.[3] Roads without milestones were not to be traveled, so milestones were invented and put in places where what was being measured was highly problematic. However, those appointed to senior positions with responsibilities for communities expected to be able to handle everything, and on those somewhat rare occasions when they asked for advice from an anthropologist they did not ask for the advice as if they did not know; they asked for advice as if they wanted to see if the advisor knew what was really happening. They found it almost impossible to admit that they did not know how to handle a problem. Managers were confident that they could overcome any technical impediment in order to get more tons out of the ground. They tended to regard those who lived in nearby communities in much the same way as they regarded employees. Why, in small towns, where much of the infrastructure—power, water, rubbish collection, and sports facilities—is owned by the company did those in town hall meetings not follow the company line? You occasionally got the feeling that they thought, but knew it would be unwise to say, that local people ought to be grateful.

Relationships were all very well, but what about numbers? If those doing environmental work could measure everything that was thought to be important why could the communities people not do the same? The accountants and engineers who run mining companies have a low tolerance for ambiguity, and they want to oversee what is to be done in the community and see evidence of progress that they can easily understand. They had a suspicion that a problem that had not been thought through.[4]

## CHOOSING SOCIAL RELATIONS

Among those with community experience within Rio Tinto there was a consensus that it would make sense to concentrate on relationships between the community and the company. Rio Tinto's decision to conduct community relations as social relations was underpinned by a belief that social relations held the key to building competence in the community. Social relationships could provide the basic infrastructure for cooperation since they could contain, provide context, transact, and give meaning to economic activity, health, welfare, and community representation. The aim of community performance should be to develop relationships between the company and the community characterized by mutual trust, mutual respect, and mutual understanding. A *Communities Policy* and a set of principles called *The Way We Work* were developed in London in 1997 to support the new science-based approach to community relations.

The new policy emphasized that good social relations were to be put at the heart of the company's performance in the community. It was essential that any policy statement should not be too long because wordiness might fail to recommend the advice to those who should use it as daily guidance, nor could the policy be so short that it failed to provide useful guidance to busy people. The wording of the policy made it clear that the relationships endorsed by the policy were to be viewed as two-way obligations. They were to be pursued from exploration through to the end of mine life; they were to be seen, as relationships must be, as aspirational because they would require continual striving on the part of the company and the community. Implementation of the policy did not need specialized education or training. Mine sites would be able to assume that an acceptable level of performance was in the process of being achieved when relationships with the community were characterized by mutual trust, mutual respect, and mutual understanding. Because universal application was the intention the policy was deliberately drafted to fit the social circumstances of developed as well as developing countries; it had to be capable of accommodating cultural diversity ranging from small pre-literate societies to large minority groups such as the Maori in New Zealand.

How should one counter the popular idea that culture could be easily changed? Senior executives talked about "changing the culture" in some part of the organization as if culture could be molded like plasticine. After it was discovered that among the almost 200 staff at headquarters there were more than 30 different nationalities senior executives began to talk about the advantages of a culturally diverse workforce. When community relations overseas are staffed by nationals it is sometimes thought that cultural factors do not need any special attention. However, in a number of instances graduates working in the community may have spent much of their lives at universities in other countries or in urban environments in their own countries, with the result that they need to immerse themselves in community life in order to gain cultural fluency. Multinational companies have benefited from the cultural differences that exist between their employees; these differences have an educational aspect in that they emphasize the importance of accounting for culture. To do otherwise would risk creating a "cultural no-man's-land,"[5] something that would happen if and when it is not possible to incorporate and make productive use of the cultural heritage of individual employees. This is always a risk when global companies try to transform Americans, Australians, Zambians, Argentineans, Englishmen, Dutchmen, Indians, Pakistanis, even Irishmen from both sides of their border into a common mold!

Within the company trust and mutual respect had to be seen to mean the same thing on the shop floor as in the boardroom and to ensure that was the case companies had to cascade a uniform understanding of key terms down through the organization. When doubters tongue-in-cheek asked what would happen to a manager who made money and brought projects in on time and under budget but paid no attention to community performance they were told that such a person could not expect to achieve a high position within Rio Tinto. Close contact with the CEO in the initial introductory period was essential. This contact sent a signal to senior managers that the company was serious about doing better in the community.

Any management arrangements for community relations had to provide for the provision of advice and supervision to sites overseas throughout the project cycle. To help senior management orchestrate the development of a new approach to community relations Rio Tinto began the improvement exercise by establishing a small team which grew from two to four well-qualified and highly experienced specialists at corporate headquarters in London. Between them the members of the team had a good mixture of mining company experience, social science expertise, hands-on community experience, and familiarity with aid agencies and NGOs. The diagnostic ability of corporate specialists appeared to owe more to instinct and facts than conventional wisdom as reflected in manuals and advisory notes. By going to sites and identifying a critical bottleneck constraint that others had not seen they could add value. Their insights might bring into play an idea from the social science literature, lessons from another site, or the need to change personnel or to try to alter community perceptions and attitudes.[6]

The corporate team was expected to provide a repository for the community relations experience which Rio Tinto had acquired around the world as well as the experiences of other miners, aid agencies, and NGOs. Oil company experience suggested that it was important to avoid starting each new project with no readily available lessons drawn from prior experience. The oil and gas industry seemed to start each project as if the company had no lessons of experience, leaving it to the local manager to decide what and how things should be done. During the early days of community relations each mine manager was allowed to develop his own idea of what should happen in the community. Not surprisingly this approach produced a wide range of outcomes and wasted resources.

Rio Tinto was decentralized and the intention was not to increase the power of the center with respect to community relations but to provide overseas sites with more guidance and support within a familiar management framework. Although the introduction of science-based community relations was initially dependent on strong CEO support, it was demand from below that, over the long term, proved to be the major factor in securing success at overseas operations.[7] Operations could be fiercely independent and often wanted to work out their own ways of dealing with challenges.

The Managing Director at an African copper mine got to his feet at the end of a dinner he had hosted and said, "There are two things we do not believe. First, those who say, I am here from London to help you, and second, the check is in the mail." It was from the beginning quite clear that if sites did not think that London could add value to what they were doing then cooperation and the chance for improvement would be limited.

The London corporate headquarters was organized into product groups dealing with iron ore, copper and gold, diamonds, coal and energy, and industrial minerals such as borates which are used in making paint and ceramics. Each of these product groups had mines and operations overseas and each had its own special social problems. These might involve community pressure for a greater local spend on economic development, trying to understand how to cope with the awful consequences of HIV/AIDS, or dealing with workers unused to managing money who were running up massive household debts.

Line managers were left in no doubt that they were responsible for comunity relations and had to get over the detail. The "I am just an engineer" excuse was no longer widely acceptable as a reason for senior managers failing to come to grips with their community relations responsibilities. Titles which suggested that communities personnel had line authority could cause confusion. For example, a title such as "Practice Leader" provided an impression of executive authority. Yet when a company patterned titles on law firms or medical practice then this would be misleading because in those situations there is no line management. The practice leaders are the rain makers and this is not the case in a mining company. Unlike the case with law or medicine, community personnel have no well-developed, or well-agreed upon, body of knowledge that can be routinely applied. There is no Law Society or General Medical Council to oversee and provide assurance about performance.

## Where in the Organization?

Where community relations was placed in the organization would send a signal. Obviously critics would maintain that a link with public relations suggested that the company viewed work in the community as a matter of window dressing requiring media skills rather than the social science skills needed to work on commnity relationships. For this and other reasons, any attempt to join communities with external relations at corporate headquarters proved to be like trying to mix oil and water. The hands-on skills of communities personnel and the media skills of external relations personnel were very different; external relations personnel gazed upward to top people and big important events, and community personnel were interested in small local issues. To maintain a strong technical identity communities people needed to work closely with the other functions that had community interests such as health, safety, and the environment (HSE). That coupling could provide a good placement but only if those in charge had the practical skill and the intellectual substance—and unfortunately that was not always the case—to forego attempts to integrate the two functions while putting in place management arrangements that encouraged, and made possible, coordination of the distinctive characteristics of both disciplines. Environmental concerns were primarily exogenous in nature; they had to do with meeting and complying with external requirements and standards imposed by international, national, or local authorities. Meeting these regulatory requirements had often allowed a mine to proceed, to extend its life, or to close.[8] What also became obvious after Bougainville was the fact that more attention had to be paid to local environmental perceptions.

At overseas sites the same placement choices as at corporate headquarters were tried and used depending on the nature of the mining and the wishes of communities. Overseas there was also on occasion an attempt to join communities' work with the security function. Integration with the security function was the least successful because experience showed that it lowered the propensity of local people to provide high-quality information because they feared it might be used against them by the security personnel.

## Fitting into the Project Cycle

Community relations had to be integrated with the way mining projects were developed. The project cycle in mining starts with exploration and may last for 20 years or more. Only one in 100 of these exploration projects is expected to yield a commercial deposit. In the event that the deposit looks promising, years may pass in proving its nature. Other requirements include drilling to assess the size and grade of the deposit, financial and engineering analysis to assess the commercial feasibility, and social analysis not only to assess the likely social consequences of a mine but also to see how the size and scale of mining can be explained to people who may have had very little, if any experience with industry. Then environmental analysis is required to see what adjustments or mitigation must be put in place. Local people may want more compensation or a larger number of small business opportunities. If the feasibility process produces technically acceptable and bankable results the project can move to the construction phase.

Mining companies typically follow a phased approach to move from initial discovery, to operations, to eventual closure. These phases form the development pathway: Studies (Concept, Order of Magnitude, Prefeasibility, Feasibility), Construction, Operations, and Closure. A project refers to a proposed mining business in the phases prior to operations. Companies set study requirements for each phase that form the basis for understanding the ore body and building a business case for investment. The project team is responsible for executing the study requirements for each phase and preparing arguments for the project to proceed (or not) to the next. Study requirements cover all aspects required to construct, operate, and close the mine: geology, metallurgy, processing, marketing, financial, legal, social, environmental, health, and safety, among others.

Commodity prices rise and fall, corporate business strategies evolve, available investment capital fluctuates, and new technologies emerge. These can affect how a company chooses to invest in the development of new ore bodies or the expansion of existing mines. A project can be on "full speed" one year and "on hold" the next.

Regardless of whether or not a company had begun to build relationships with the community it was difficult to begin building schools and hospitals before a company knew it had a viable project. Mining companies were reluctant to make firm commitments to provide extensive community benefits if they feared that mining outcomes and market conditions were uncertain. But there was a case to be made for a neighborly relationship with communities even if a project did not proceed beyond exploration. Companies helped with first aid, medicines, evacuation to nearby clinics, and hospitals. Often bridges were built and water supplies improved because of exploration activities. On the downside companies operating in areas where government services were thin, poorly developed, or of recent provenance had learned the hard way that it was much easier to develop wants than it was to provide the means of their satisfaction.

Community specialists began to believe that there could be advantages from a new project development sequence that began with making the social case for mining. They noticed that fieldwork during exploration enabled geologists and local people to get to know each other and, like anthropological fieldworkers, they learned how to conduct relationships with informants. No promises of material aid or jobs were made, as already mentioned, because only one in 100 exploration projects would produce a mine. Exploration lasted longer than anthropological fieldwork and generated a large number of community contacts because 100 or more company personnel may have been involved. These social relationships reflected local beliefs, values, and attitudes, and they established a degree of mutual trust, understanding, and respect.

It would make sense to accumulate rather than dissipate local knowledge and relationships at each stage of the project cycle. Couldn't continuity be secured by having those who dealt with the community stay in place for as long as possible throughout the project cycle? They would gradually build local knowledge as well as mutual trust and understanding between the company and the community. Asking local people to invest their time in working at a mine or trying to improve their incomes or health is often quite risky. It takes a lot of trust to place one's own future and the future of one's family and community in the hands of people one does not know.

Steps were taken to integrate social thinking within the project cycle. Historically, companies began by making the business case because it was thought that the process of gaining acceptance should begin by providing a shop-window view of the goodies that communities, regions, and the nation could expect from mining. Benefits were phrased in terms of environmental protection, jobs, small business opportunities, local revenue generation, new infrastructure in the form of road and rail, hospitals, new health facilities, and schools. When the find was in a developing country the mine development team had to try to put themselves in the shoes of local people (whom they did not know) in order to advise management how to convince local leaders that what was on offer would be of benefit to all. The exploration team went to look for minerals somewhere else. Contact established with local leaders was expected to help foster mutual understanding, and trust, by showing willingness to improve the local quality of life.

When a company did not have an exploration-to-closure view of relationships with communities the benefit that could come from an early start with community relationships was lost as the project moved from exploration to construction. This situation occurred because, while it was the case that more and more companies included community relations in the exploration stage, too few of those companies took steps to ensure that there was continuity of personnel and contact with communities as the project moved into the operating stage. This meant that contact with communities was lost at precisely the time when community concerns arose and needed to be quickly resolved by consultation with company personnel whom the community had come to know and trust. Instead of carrying on with community contact and building up their community cadre some companies waited until construction had finished and commissioning had taken place in order to continue to work closely with their communities. Experience suggests, if we use the analogy of a railway station, that when carriages reach the station exploration personnel should not give up their seats to construction personnel, who in turn should not give up their seats to operating personnel at the next station. Some should ride all the way, thus becoming well-known to the community while advising company personnel who are merely in transit. Communities need to have the reassuring presence of company personnel whom they can get to know and trust, and this should start at the earliest possible opportunity.

Where community relations have been well-established early on in the project cycle there is a good chance that by the time community consent is needed mutual understanding and trust will have been established. Where community relations have not been established before community consent is required, the process of educating communities about the benefits and burdens of mining, and the process of trying to persuade communities that the advantages of proceeding will be greater than not investing in the project, have inevitably had to focus on offers of material help, jobs, community investment, supplier chain possibilities, and so on. When the company is focused on the management of expectations and the community is focused on leveraging benefits, disappointment on both sides is inevitable.

Construction, which can last for several years, produces its own unique social problems and opportunities. Thousands of single men may descend on a remote location. This produces substantial social problems. What is to be done with workers who may have HIV/AIDS in a country which does not permit testing? Local people, instead of continuing their relationships with geologists, know that once construction starts they have to begin to get to know a totally different cast of characters. When the mine becomes operational, new social issues emerge. These may concern the workforce, the operation of the plant, noise, dust, and so on. More difficult social problems emerge when most employees follow a roster which requires on-site work for 10 or 14 days a month followed by flying out to their homes for the rest of the month. What needs to be kept in mind is that a mining project, quite unlike a development assistance project run by an aid agency, can last for 50 or 100 years. Community focus in such cases is almost always on compensation and jobs—compensation fades quickly after construction ends, and skilled jobs disappear, then people start to evaluate the social impacts.

## FIVE-YEAR PLANS

The requirement that each overseas operation should develop a five-year plan was intended to emphasize the need to carefully consider what community relations should deliver for the business. An annual review was not sufficient to get a sense of progress (or a lack of it), since some activities lasted for several years and others needed lengthy preparation. Operations came up with new investment and activity ideas each year but without at the same time delivering any good assessment of what had happened to the previous year's bright new initiatives. Normal site reporting did not cover the sorts of information involved in the five-year plan process, and sites were often reluctant to tell corporate headquarters that they had any problems at all.

What plan period should be selected? Plans in developing countries were often for five or seven years because these plans were primarily concerned with agricultural development, wherein much time must pass before tree crops reached production. A rolling five-year plan was selected by Rio Tinto as a way of improving the overall management of community relations. Of course, five years was not necessarily appropriate for all operations; some, because of a need to secure a time-bound result, might ask for, and be given, a shorter or longer period.

The process whereby plans were put together was as important as the content. It depended on parties that might not have worked closely together (the managing director, the communities team, and the community) getting to know one another and starting a contact process that should last as long as the plan which they would jointly own and implement. The process began with the Managing Director saying what the business needed over the plan period and then community wishes could be discussed and added. Working on the plan ensured that mine managers knew what communities wanted.

How those business goals were achieved would help the business and might well make a contribution to poverty alleviation and other international objectives. But an operation that was too focused on aid agency objectives was often not sufficiently focused on what the function could do to minimize project delays or expensive local disagreements.

The plan established relationships between corporate specialists and communities teams and made site personnel aware of the range and quality of advice and assistance that they could receive from the corporate center. The plan produced perspectives which enabled both the site and the corporate center to plan collaboration on the basis of where they were, what sort of progress was being made, and where extra effort was needed. At the corporate center the plans could be looked at in the whole to provide a picture of the health of the function throughout the company that could be used to respond to critics, provide material for annual reports, or to brief those in charge of geographical regions or product groups such as energy or diamonds, iron ore, or copper to get a sense of how well their sites were doing.

The five-year plan provided the *raison d'être* for visits by corporate specialists to the site. Five-year plans created and maintained personal relationships between sites and the corporate headquarters; it was a two-way relationship which enabled sites to provide opinions, progress reports, queries and requests on a semi-official basis and enabled corporate specialists to provide advice and support thereby spreading their technical skills over a wide area. The presence of engineers and technical personnel in the planning process helped to make sure that a few high-priority bottleneck problems were addressed rather than spreading resources too widely and too thinly to make a significant impact.

In the five-year plan sites were expected to look at the gap between where the operation was in terms of its local relationships and where those relationships needed to be in five years. Plan objectives might involve getting community approval for an extension, greater involvement in the business by local residents, or responses to local economic development requests. What improvement would be necessary and what resources would be required and what needs to be done to achieve the changes that were wanted? The plan was rolled over year by year in a review process which examined what had been achieved and what was planned for the next year.

## Hiring

At the community level the number of employees varied from country to country and from company to company. A Rio Tinto staff of 20 dealt with a community of 2000 for an operation producing titanium in Madagascar, and a South African miner used a staff of 5 to handle 100,000 in a platinum operation. In Indonesia a miner used a staff of 100 to deal with 30,000. Hiring for community relations was often a matter of self-selection: those who had lived and worked in a community for a number of years decided that they no longer wanted to work as a geologist or engineer. While this method produced candidates who had already demonstrated that they could get on with local people if continued, it did not do enough to raise the skills and knowledge of those who worked in the community so that they could deliver what critics of the industry's community performance were demanding.

In order to hire senior social science specialists it was always thought that this needed to be led by someone with a tertiary degree and strong experience who would engage in a selection process to find similarly qualified specialists in a process that was open and transparent. The initial hiring of headquarters senior specialists had to set the tone for the future and it was felt to be important that the hiring process involved all the existing specialists. An attempt was also made to secure candidates who had the sort of regional experience that matched existing company operations. Here practice departed from what happened with the other mining disciplines such as geology and where hiring could end up with a manager who did not have a strong technical background. If the hiring of community specialists was given to a single manager who lacked both strong social science credentials and the good sense to ask for assistance the results might be very expensive. Social science–oriented CVs can be mendacious documents. Would-be candidates produced beautiful CVs indicating that they had held senior appointments since the age of seven or ten. Care had to be exercised to make sure that entering professionals had the right attitude and the right experience. Too many wanted start at the top or as near to the top as they could as assistants, advisors, or bag carriers for the top people. They wanted to start by working on policy and other big issues. Hiring for senior specialist positions at the headquarters was done by committee because of the need to consider a range of factors that experience had demonstrated were unlikely to be possessed by a single manager. To be a useful candidate for a permanent headquarters advisory post an applicant needed to have had a good tertiary degree in a relevant subject and should have shown that he or she could make a difference at site level. Such a person had to be sufficiently abreast with what was happening in the research field to be able to identify problems and know where to go and who to ask outside the company for any help that might be needed; he or she needed to be able to write tight terms of reference; needed to be capable of supervising the work as it progressed to ensure value for money; and needed to be able to ensure that the work was effectively used. Such persons would be of little use if they were merely specialists in a particular discipline or profession. They had to understand the way a large company operated and had to possess minimal bureaucratic competence as well as an ability to communicate effectively and economically with the most senior personnel in the company.

## GENERALISTS AND SPECIALISTS

Company staffing has to decide on the mix of generalists they need in the field and specialists they need at corporate headquarters. Though both are important to the success of the operation, the roles and responsibilities of the specialists and generalists vary appreciably. The specialists have skills that are relatively scarce, are exercised by them personally, and constitute the basis of their influence. The generalists have less identifiable skills, work through others, and secure their influence because of their ability to provide integrity to the system as a whole. Generally speaking, the contributions of the specialist are not bound by institutional constraints and are limited typically by his or her own competence and capacity to convince others as to their quality. Generalists are far more constrained by the institution of which they are a part; much of their expertise rests on their understanding of that operation and those related to it.

While a company needs its social specialists, it is also true that many skills will, at best, be underutilized. Many of the specialists are involved less in the practice of their skills and more in leadership tasks, such as influencing others in their organization, and in other organizations, to follow different ways of behaving, selling programs, struggling for resources, relating with other sectors on joint programs, and so forth. For those in operations who really do practice a specialty, opportunities for development will primarily lie in the professional discipline and this will probably mean periodic returns to a university. In contrast, the generalist requires the development of an ever-widening perspective. Instead of returning to an institution for retooling, the need is for change and the development of new perspectives both with respect to their own operation and those that are part of its environment. Such broadening may occur in many different contexts; a different job, a special assignment, a training program designed to meet their generalist needs, or the university.

## HANDS-ON IN THE COMMUNITY

Prior to Bougainville community relations was a matter of what mining companies did *to* their neighbors rather than what they did *with* them. Community relations were conducted as a series of one-way rather than two-way relationships. Initially the mining and community relationship was driven by philanthropy because companies assumed that conspicuous acts of generosity in the community—building schools and hospitals—would induce local people to like miners. That did not happen. Giving was succeeded by economic realism and an assumption that what really counted was the money in the pockets of local people or spreading entrepreneurial opportunities as widely as possible among the neighbors of the mine.

It was wise to be careful when tempted to do good deeds. A colleague responded to a request for a few bags of cement to help with the construction of a statue which he was told would help the fledgling tourist industry in northeast Brazil where the mine was located. When the company was asked to attend the unveiling of the statue he went along with his camera. Certainly the statue looked impressive from the back as they climbed to where it stood on top of the hill. In the town below they could see it was well made but were puzzled as to why, instead of a likeness of Christ, the face was that of the mayor who was running for re-election.

## Community Relations or Stakeholder Relations?

Mining companies that have not developed good community understanding and hands-on community skills sometimes rely on stakeholder rather than community relations. Stakeholder relations seems a misnomer because it is about individuals rather than community relationships, and it is hard not to escape the conclusion that this is a form of analysis more suited to situations in the industrialized countries than to remote areas.[9] NGO stakeholders are seen as persons who need to be consulted because it is believed that they have interests in a business that needs to be considered. Stakeholders used to be defined as persons who held the wager money committed by two or more persons who were competing in a bet or sporting event. Now stakeholders are those whose opinions about a mining company are considered important by those conducting the analysis. A local community is not itself seen as a stakeholder because this form of analysis cannot handle collective representations or relationships.[10]

How can a mining company which has no baseline data, and is operating in a remote region where government services are of recent provenance, decide who is an opinion leader in the local community? Can a company which has not engaged with the community look at a social landscape and work out who is an opinion leader? Church leaders? Tribal leaders? Politicians? Is there no need to have any confirmation from the community that the stakeholder identification exercise is accurate?

Who in the mining company will design a statistically valid sampling strategy? Stakeholder relations relies on opinion survey methods and techniques and companies that utilize it extensively will have to develop an unusual degree of statistical competence. Communities (and other company) personnel doing stakeholder analysis on behalf of mining companies seldom have the skills to rigorously define samples and sample sizes and nor do they have the skills to analyze the responses. What happens is that the samples seldom reflect social reality either because status and social stratification have not been understood or the manner in which community opinion is formed and communicated in the aggregate has not been researched. Small sample sizes and low questionnaire response rates are common and are bread and butter issues for opinion survey firms working in the rich countries but pose difficulties for miners working in remote communities. Survey responses may be tabulated without sufficient attention being paid to who has responded and why they have responded. This is important because it is often the case that those most passionate in their support for or opposition to what a company is doing or proposes to do will be the most frequent respondents.[11]

Is there evidence to show that stakeholder work has shifted opinion in remote communities? Can proponents show that attitudes and opinions in the community have shifted or are changing as a result of company messaging suggested by stakeholder work? The consulting company Environmental Resources Management (ERM)[12] recently said: "activities under the banner of 'sustainable development' (e.g. stakeholder relations, CSR, social and environmental performance) tend to be lauded as key components of many firms' risk management strategies. In reality, as currently designed, many of these activities fail to add value and they provide many companies with a false sense of security. The current model has reached its limits and it's time for a change."

The administration of a stakeholder questionnaire is not a conversation because there is no exchange of information, no opportunity to ask about a sick relative or the performance of a child at school, no opportunity to inject warmth or concern into the process. Those who administer the questionnaire will not say, "see you next week," they will not go to a funeral, or a marriage or a feast. The questionnaire is not an exchange; it reflects what the company wants to know but may do little to reflect what the community wants to know; it will reflect company priorities and not necessarily community priorities.

What will be the motivation behind a response? This is not a question that is frequently posed and is certainly seldom answered. Stakeholder analysis is not sufficiently reciprocal, fine-grained, or nimble to provide the constant flow of information as well as the changing perspectives that would be necessary to service community relations.[13] Continuous surveying would be required and that would result in low-quality data and questionnaire fatigue. The more ambitious stakeholder relations becomes, the more it will encounter increasing complexity and differentiation as it has to survey more and more individuals who have different points of view. Size matters: Graicunas's Law holds that in any organization the number of individuals will increase arithmetically as one goes down through the levels of the organization but the number of relationships will increase geometrically.[14]

For those who have grown up in an industrialized country with surveys and polls the idea of stakeholder relations is not hard to accept. Yet the experience of people living in remote communities may be very different. What do those who are stakeholders think? Do their views and opinions come close to those of the surveyors? Stakeholder analysis cannot assume the existence of understanding about the nature and purpose of this way of looking at information-gathering and messaging. The use of this model is inevitably deductive and effort is needed to show the local meanings. Social relationships persist through time and require reciprocity, whereas survey relationships are one-off events and those surveyed have no sense of obligation with respect to the reciprocity normally present in social relationships. Nevertheless a number of companies have made stakeholder relations a component of their community relations, believing that both functions cover somewhat the same ground. They do not. The fact is, and this has not been grasped by a number of mining companies, stakeholder relations are closer to public relations than community relations and are best at addressing reputational questions that are of interest to media personnel.

Stakeholder relations will make most sense in countries where individualism is well-advanced and media are well-developed. In remote communities, however, the role of media may be spotty or nonexistent, and opinions may be those of a tribe or an ethnic group. In Europe or North America the individual votes, is employed as an individual, and forms, holds, and expresses opinions that often differ from those of family members and neighbors. Trade union membership and religion may offer slight variations to this pattern. However, this is not necessarily the pattern that is found in developing countries where the view of the individual may reflect kinship affiliation or community solidarity. A great deal of preparatory work would have to be done to understand how local opinion is formed and changed with respect to the matters that are of interest.

## Notes

1. SOPs have a lineage that goes back to the "scientific management" thinking of Frederick Taylor which concentrated on the shop floor. Following an engineering model which supposed that how organizations were designed radically affected work performance Taylor focused on how output could be increased. Taylor tried to improve the use of the shovel in factories by redesigning the spade, rearranging the materials, and introducing new monetary incentives. See Taylor (1919).
2. See on these issues, Kemp (2006).
3. Aid agency project tools such as log frame analysis or means-ends analysis are not regularly used in the mining industry.
4. A counter argument to this somewhat stereotypical view can be found in Vaughn (1996).
5. "Cultural no-man's-land" is a term used by Malinowski in his article "Practical Anthropology" in (1929).
6. Michael Polyani calls this "tacit knowledge" which can only be detected by its action. See Polyani (1958).
7. The introduction of community relations was very much a team effort. Rio Tinto's Main Board Director Lord Holme established a working group to introduce new environmental policy capacity, new approaches to international relations and new approaches to community relations. The environmental approach was developed by Tom Burke, CBE, the international relations approach by Robert Court, and community relations by the author. Technical implementation of the community relations proposals involved personnel from the Americas, Australia and Asia, George Littlewood, John Hughes, John Senior, Preston Chiaro, Sergio Visconti, Paul Wand, David Godfrey OBE, Louis Cononelos and Alexis Fernandez. On the development of engagement thinking see Brereton and Harvey (2005).
8. Many well-known consulting companies that were initially staffed to deal with environmental issues used these environmental specialists to do social performance work as demand in the community began to ramp up to meet societal expectations. They held onto this staffing pattern because of the growing demand for environmental and social impact analysis.

9. See Freeman et al. (2007).
10. See Edwards (2000).
11. See the elementary sampling material in Cochrane (1979).
12. See Cattano (2009).
13. See Fassin (2008, 2009, 2010).
14. See Graicunas (1937).

## BIBLIOGRAPHY

Brereton, David, and Bruce Harvey. 2005. Emerging Models of Community Engagement in the Australian Minerals Industry. A paper presented at the International Conference on Engaging Communities, An Initiative of the United Nations and the Queensland Government, Brisbane, August 14–17.

Cattano, Ben. 2009. *The New Politics of Natural Resources: Time for Extractive Industries to Address Above-Ground Performance.* London: Environmental Resources Management.

Cochrane, Glynn. 1979. *The Cultural Appraisal of Development Projects*, 78–91. New York: Praeger.

Edwards, Michael. 2000. *NGO Rights and Responsibilities, A New Deal for Global Governance.* London: The Foreign Policy Center.

Fassin, Yves. 2008. SMEs and the Fallacy of Formalising CSR. *Business Ethics: A European Review* 17(4): 364–378.

———. 2009. The Stakeholder Model Refined. *Journal of Business Ethics* 84(1): 113–135.

———. 2010. A Dynamic Perspective in Freeman's Stakeholder Model. *Journal of Business Ethics* 96(1): 39–49.

Freeman, R., Jeffrey Harrison, and Andrew C. Wicks. 2007. *Managing for Stakeholders: Survival, Reputation and Success.* New Haven: Yale University Press.

Graicunas, V.A. 1937. Relationships in Organizations. In *Papers on the Science of Administration*, ed. Luther Gulick and F. Urwick Lyndall. New York: Colombia University Institute of Public Management.

Kemp, Deanna. 2006. Between a Rock and a Hard Place: Community Relations Work in the Minerals Industry, PhD thesis, University of Queensland.

Malinowski, Bronisław. 1929. Practical Anthropology. *Africa* 2(1): 22–38.

Polyani, M. 1958. *Personal Knowledge: Toward a Post-critical Philosophy*, 50. Chicago: University of Chicago Press.

Taylor, Frederick W. 1919. *The Principles of Scientific Management.* New York: Harpers.

Vaughn, Diane. 1996. *The Challenger Launch Decision: Risky Technology, Culture, and Deviance at NASA.* Chicago: University of Chicago Press.

# CHAPTER 6

# A Systematic Approach

Rio Tinto's community relations consultations and project experience suggested that three fundamental stages should be followed in sequence in the approach to community relations. The social baseline was the first essential stage in building the foundation of local knowledge and understanding that was required to design and handle social relationships with communities and in understanding the way communities viewed the company. A good baseline could provide an indication of what community relations had to do over the next five or ten years to achieve its objectives and in so doing provided an indication of the skill sets that would be required.

Using the information generated by the baseline the second stage of the approach to community relations was to understand, record, and make available to all working in the community a detailed description of the way in which communities handled consultation and decision-making. A written record, agreed with the community, should be distributed to all who were expected to come in contact with the community. This helped to verify and authenticate the process of consultation and negotiation for third parties. In all three stages emphasis was placed on the importance of proper record keeping. The mining industry does badly in failing to keep records of simple meetings, commitments, and understandings, and this combined with rapid turnover of personnel contributes to much of the misunderstanding that underpins much of the friction between community and company.

Using the information generated by the first two stages of the approach the third stage was to assess those activities, economic or otherwise, where the company and the community could work in partnership to achieve a

© The Author(s) 2017
G. Cochrane, *Anthropology in the Mining Industry*,
DOI 10.1007/978-3-319-50310-3_6

result that would be of mutual benefit. Companies had learned the hard way that when operating in remote areas it was all too easy to take over the role that the government should be playing and, once this happened, it was usually the case that it became extremely difficult to return to a more modest supporting role.

To avoid giving the impression that unlimited company assistance would be provided companies needed to say just who they thought their neighbors were and where they lived. African experience suggested that those living within 50 kilometers of the operation could be considered to be neighbors who would be covered by the intention to provide jobs and opportunities for economic development.

Local investment was not a matter of trying to find out what might be done to improve local welfare; it was also a matter of knowing what cannot be done, what should not be attempted. If those who want to recommend investment cannot identify areas and activities that should not be handled, then a great deal of money may be wasted with little return.

## THE SOCIAL BASELINE

A baseline was not intended to be seen as a series of hoops through which it must jump; rather it should be seen as real features of the terrain across which the company would operate. Just as a road's design must be suited to the physical terrain, so must many relationships be adjusted to the particular social terrain, some features of which would pose major design questions while other features may not affect the activity or, indeed, may be counted upon to significantly boost the probability of success.[1] The baseline was not a matter of developing their understanding of us and our understanding of them; it was instead a process of blending and mingling these two views.

A collaborative effort was required involving headquarters specialists working with staff at operations. Traditional anthropological fieldwork could produce data of quality but collection took a lot of time and the coverage might be quite narrow. A company that was serious about its community relations could not risk a process that might produce either superficial knowledge or endless debate and disagreement. The anthropologists at corporate headquarters in collaboration with the communities team at the operation assessed what was known, what was needed, and how understanding was to be updated. The communities personnel were not always professional anthropologists and could not be expected to perform in the same way as professional anthropologists but they knew the local peoples and knew what information was needed. Where the local team did not have the skills they worked with locally based anthropological consultants and the anthropologists located at corporate headquarters.

Investigation and elicitation could begin with the history, contemporary beliefs and values, ritual practices, kinship, and the world view that animate social organization. That would include rites of passage, birth, coming of age, marriage, and death; beliefs, values, attitudes, and sanctions; sacred and profane typologies; ceremonial life; the nature and definition of power and status; and social structure, leadership, and seniority. But investigation needed to keep an eye on some essential questions. What did trust look like? Trust existed when the information obtained in community conversations is information that might lead to the informant suffering embarrassment or ridicule or even prosecution. Trust existed when local people allowed a company to move graves from one location to another to accommodate construction needs. The easiest way to lose trust was to make promises that could not be met, promises that may well have been made with the best of intentions. How could trust, mutual understanding, and respect be earned and what was needed to ensure that, once gained, the good opinion of the community was kept?

What about factionalism? Would the community work well together? When there were strong factions in a community, that might impede communication and cooperation. Morris Carstairs showed how difficult and complex it can be to learn when, as a doctor in India, he had to move between different castes to serve his patients.[2] What processes and procedures did the community have to handle social control? Where baseline description and analysis were dealing with communities where custom and tradition is strong it should have included concepts relating to deviance and wrongdoing, collective, individual and spiritual agency, real and personal property, dispute resolution, and positive and negative sanctions.

If not available a cost-of-living index should be constructed at the earliest possible moment and should track price movements in a basket of local essentials such as rice, cooking oil, meat and chicken, school fees, and clothing prices, and, where relevant, house prices. In many instances mining can quite dramatically increase the cost of basic foodstuffs for urban residents who are not employed by the mining company. Hoarding and monopolistic positions should be identified and corrective action adopted.

Health was a major concern. Sexually transmitted diseases were a concern at most operations beginning with construction, and as a result companies have become heavily involved in awareness programs. The challenge may be to deal with the orphans and grandparents as the only surviving family members after parents have succumbed to HIV/AIDS. Respiratory diseases and in particular the nexus between TB and HIV/AIDS needed to be continuously monitored.

Famine and malnutrition was also important. In areas where there is food scarcity a careful record should be kept of amounts and types of food retained from yields and used for family consumption or reserve. This record should include livestock as well as grains, tubers, fruits, and other vegetables, and tree crops. In addition, it is useful to record the most common methods of food preparation and actual food distribution within the family. What criteria are used for intra-family food distribution? Do males get more than females? Within the community itself, food given in exchange for goods and services enjoyed by the family should also be recorded. How much goes to the landlord or other families for satisfaction of debts contracted with stores and so on? The portion of family production offered for sale should be recorded; sometimes farmers are forced to sell all their marketable surplus when prices are low, sometimes they have poor leverage and must sell individually to well-organized middlemen, and sometimes they sell only when cash is required. Nutrition and mother-and-child welfare can become worrying in situations where workers from areas where they have been dependent on subsistence agriculture suddenly end up as wage earners having to buy food instead of growing it—their initial food basket choices are not always optimal.

Investigation of local tenure could turn up surprising facts. A baseline in Indonesian Papua helped Freeport to understand that a local tribe's claim to land ownership might not be as ancient as had been claimed. The tribe told the mining company that they had possessed land around the mine since time immemorial. Proof of land ownership normally required having 10 to 13 generations of ancestors living on the area without interruption. In this instance kinship and genealogical research showed that those making the claim only had a genealogical depth of three or four generations; they were recent arrivals rather than time immemorial owners.

Baselines produced other information that was not always obvious. Were local people sitters or squatters and did male and female toilet facilities have to be located some distance from each other because of avoidance rituals? Should employees live in company accommodation or in the community? Did children of a certain age need their own bedroom? When designing housing should kitchens be separated from sleeping quarters?

The local social knowledge that a baseline produced should be able to help when planning health and safety issues, for example, to judge the risks of operating heavy equipment during Ramadan. What should a company do to make provision for time off from work for funerals and locally important rituals? Strategies were also required for hiring during times of ethnic tensions between local groups. Family breakdown could be caused by fly-in and fly-out work rosters where workers leave their homes for four or six weeks work at a mine and then go home for a couple of weeks—this separation strained marriages and could deprive children of both their parents at key times when they are growing up. How should poor performance be punished? Dressing down local personnel who performed poorly could have negative consequences unless the local disciplinary norms were understood and admonition was administered in a culturally appropriate way.

Since industrial development accelerates the speed of social change, mining companies have had a special responsibility to identify and understand the nature of social changes attributable to the operation, as well as the nature of the changes that might be taking place in local society regardless of the presence of mining. As already mentioned in relation to indigenous peoples care has to be exercised to separate out those who, in the opinion of analysts, would continue to proceed regardless of the presence or absence of the operation. This calculation and recording is necessary because companies are often blamed for changes that would have taken place regardless of whether or not there was mining. Because mining can accelerate the process of social change companies have a special responsibility to try to understand what these changes would look like, when they might take place, and what could be done to soften their impact.[3]

## THE PENAN BASELINE

The Penan of Borneo are an iconic forest people whose way of life was changing rapidly as a result of the government of Sarawak's dam construction ambitions. There were sharp divisions of opinion about what change might mean for the Penan. When the author visited Penan living in the vicinity of the proposed Murum dam in January 2006 the life of the Penan that had been recorded by Professor Rodney Needham 40 years previously no longer existed.[4] Logging roads were rapidly being built all the way to the Indonesian border.[5] Evidence of extensive logging was everywhere—rivers had been polluted and game had disappeared. Such had been the speed of change—or "development" as Sarawak politicians saw it—that the clock could no longer be turned back. The Sarawak government and the rich developers had moved on to plantation agriculture and wanted the cheap labor represented by the indigenous peoples to help produce oil palm and pepper.

To better understand what changes were inevitable, what change was induced, and what could be done to do better with change management a social baseline was undertaken by Rio Tinto and presented to the government of Sarawak (when provided to the government it was termed a contemporary ethnography).[6] The 100,000 Dayak who lived in the forests of the Sarawak portion of Central Borneo at the beginning of the last century had been reduced to little over 1000 comprising groups of Iban, Bidayuh, Kayan, Kedayan, Murut, Punan, Bisayah, Kelabit, and Penan.

A small team of anthropologists and sociologists who had worked in the area was recruited. Agricultural, health, and business specialists were also brought into the team as well as wildlife experts to undertake game surveys designed to understand how animal movements and habitats had been affected by logging. The team spent the next year collecting data while working in the field visiting the various Longhouses in the Murum catchment area. The results were discussed with local NGO groups as well as Cultural Survival and the government of Sarawak. Not surprisingly, the team's conclusion was that resettlement should take place within the catchment area. We believed that logging would have to be curtailed because any dam would need catchment protection that only a forested area could provide. The baseline suggested that a forest reserve should be created and regulated. Within this forest area the Penan could pursue traditional pursuits albeit not with traditional efficiency. Since logging activity commenced in the catchment, available forest resources have dwindled, and yields gained from hunting and gathering have declined markedly in relation to the effort expended in pursuing such activity. For example, hunters continue to search for game with guns and blowpipes but have to travel greater distances to secure a catch. As a result, Murum Penan communities now rely more heavily on cultivated crops for subsistence and for sale and on employment. These livelihood activities are supplemented by hunting and gathering, which remains economically and culturally important to Penan. Rivers remain central to transport, but roads and logging tracks have now also become key to transport, and are important features of the lived environment in the Murum catchment.

The really tricky question was what the mix of traditional pursuits and smallholder agriculture should be in terms of plots and access. Given the way Penan had adapted to living close to the loggers it appeared likely that there would be a gradual shift from traditional pursuits to cash cropping but the speed at which such a shift could take place could not be predicted with accuracy. The government wanted to accelerate the change by only providing land that could be used for subsistence and cash cropping whereas the baseline indicated that any change should take place at a speed which was dictated by the Penan themselves.

How Penan allocated their time to food collection and money-making had been affected by their living in permanent houses donated by timber companies. This meant that the Penan were anchored in one place and less able to move as their environment was gradually degraded and government health and education services were located much further from their Longhouse. A degraded environment forced the Penan to spend more time than was the case in the past in their search for game, harvesting activities in the jungle, fishing, working to get money, and traveling long distances to get health services or education for their children.

Rivers are roads for the Penan and consequently are central to life in the Murum catchment. They have been and remain essential transport routes, but equally importantly, they form the framework within which Penan organize their understandings of their social life in its environmental context. They enable travel to forest sites stocked with resources and memories: plants under stewardship are accessible because of them and historical and heritage sites are located along them. Rivers are therefore repositories of cultural and environmental knowledge in themselves and collectively, and Murum Penan possess an encyclopedic knowledge of the Danum, Plieran, and their countless tributaries. They are aware of a great variety of river features, such as pools, rapids, and minor streams, and refer to them in terms of their own social history.

Other features, such as the community's footprints across the landscape, are rather less perceptible to the untrained eye. These include burial sites, former settlements, botanical features under *molong*, or rivers and pools which bear names relating to the community's migration history. These sites are nevertheless historical records as well as location markers: they hold Murum Penan social memory relating to their origins and routes across the landscape.

After dam construction and resettlement many of these sites would be submerged, and with them, the record they collectively hold of Murum

Penan history. Erasing this record from the landscape could sever the social link between individuals' memories and the mnemonic landscape. A new form of social repository for Murum Penan history is required, and this should take a multimedia approach to mimic as closely as possible the experience of a narrated landscape. This repository could consist of oral history narratives recorded while moving through the landscape, accompanied by maps, charts, place markers, and documentary records.

## THE DOUBLE AGENT PHENOMENON

The job in the community could be thought of as being like that of a double agent because those who work in the local community occupy a halfway house—they need to avoid being seen as favoring either the community or the company. Their community performance has to represent both points of view in a credible way. Those who work for the company in the community or who try to represent the community in the company serve two masters who may strongly disagree about what is to be done and how things are to be done.[7] Good community relations people act as advocates for the community which can place them at odds with management which expects them to place the company interest first in all things. "Agent" is a nimble term with a long history, and it has the advantage of seeming to provide for the views of both company and community—one major company, Freeport, uses "Community Liaison Officer" in its Indonesian operation. This has obvious appeal. It is always necessary to keep in mind the need to consider how to motivate those working in a communities team. The job would never pay very well and those who were in it would be unlikely to rise to senior positions in the company. Few would be transferred to other operations in other countries because their local knowledge and contacts would be seen as invaluable to the operation. Obviously, working in a community could entail considerable stresses and strains and could be a lonely existence. It cannot be assumed that relationships will be governed by the values, beliefs, and attitudes of the mining community—community members do not necessarily work for the company and relationships should reflect this reality. Communities personnel rely on personally earned authority. This is an asset that comes from the way relationships with a wide variety of individuals have been formed, nourished, and handled day to day. An engineer can hand over her or his responsibilities to another but this is not the case with those working in the community because those relationships cannot be passed on to another with any assurance that the transfer will create continuity. Those working in the community have a liminal existence: neither fully and completely a part of the commercial and operational part of the company nor fully integrated into the community.

Those who want to do well in communities should not begin by thinking of themselves as being superior to local people. Spend time in any community that is in regular contact with a mining company and you will discover that local people have little time for those from the mine who come round only when they need something. The job needs agility, creativity, and subtlety if the incumbent is to serve his or her community masters as well as those who must be obeyed in the mining company without arousing suspicion and unhappiness in either camp.[8]

It helps to be seen as a reliable confidant because it is then possible to have a good sense of how a community will react to a proposal, whose views will carry weight, and whose will not. What about the motivations behind a community decision? How strong is opinion for and against? Any notion of super-ordination and subordination is not really appropriate for community agents because communities do not work for companies. Communities are not patients or clients, and those who work in communities are not doctors or lawyers. Charles Erasmus told of an extension agent in South America who used to turn up in his communities in a pinstripe suit and Homburg hat[9]; fitting in is usually more important than standing out. Relationships do have to be fixed and maintained in good working order and this requires rolled-up sleeves and getting hands dirty.

Many communities are more willing to consider an idea that comes from a trusted person rather than a solution which comes from an impersonal source. Doing small things well over time is what counts. Substantial social change needs lasting relationships with helpers who have the capacity and the willingness to not only live close by or in the communities they seek to assist but to also have the language and hands-on skills to earn local understanding and respect. When rapport has been established sensitive information may appear. An example of this is the existence of dry sex, a process that has been found in Africa and in Indonesia. It involves a woman drying all vaginal secretions before sexual intercourse in order to provide more pleasure to a man than conventional intercourse but as a result increasing the risk of spreading HIV/AIDS.

Having the skills of a handyman—a man the French call a *bricoleur*—will help because it is continually necessary to fashion useful things out of odd bits and pieces of happenings and information. When the formal banking system could not cope with small loans at a South African mine, communities personnel came up with the idea of establishing a sidewalk bank near their Richard's Bay mine to provide women with the money they needed to trade on a daily basis in the market. The premises consisted of a freight container that had been turned into two rooms. In the morning funds, for which a service charge was levied, were handed out, and in the evening the money was repaid. Borrowers and lenders knew each other, the default rate was zero. Once, when a thief tried to steal from the till, women traders caught him and handed him over to the police. In Vermont, USA, a helicopter pad, constructed by an operation that produces talcum powder, aimed to assist with medical evacuation. Within a month of completion the pad had helped to save a life. Practitioners came up with solutions to fit local social circumstances, and they did so quickly and with a minimum of fuss. However, when they had a working solution they did not hawk it round the world to try to get all other operations to adopt it.

To work diligently with a range of relationships is what communities work is all about at an overseas operation. The stuff of neighborliness may involve participating and contributing to events in social and ritual calendar events, providing personal help with gardening, fishing, or hunting, sympathizing over the burdens created by unruly or disappointing children, engaging in community fundraising and sports, cluck-clucking over disputes with neighbors, and offering support with illness or misfortune. Performance creates a sense of obligation on the part of the recipients; such an obligation is an understanding, not as strong as an agreement, but in many significant ways more binding on the parties. At a time of his or her choosing the practitioner is able to call in and mobilize all these different obligations so that they constitute a platform which supports and underpins existing relationships with the mining company or generates support for changes to that relationship.

Good social relationships have been developed by companies identifying, training, motivating, and supervising their personnel in communities. Without the right stuff in place there is little point in assuming that aid agency reports, league tables, and guidance manuals would do the job. Good relationships are developed by doing the hard yards, day by day, relationship by relationship, activity by activity, conversation by conversation, cup of tea by cup of tea. Resonance came from telling jokes and involvement in the ongoing, engrossing happenings of local society. Participation is essential in funerals large and small, and in fetes and parties. Those who manage field personnel must do their best to try to make sure that the way they work is understood by mine management since so much of their effort may not be captured by formal reporting.

Community relations around the world tend to take on some of the unique characteristics of the regions where they are located. In North America there is a strong push to educate the public about mining, and this is combined with a desire to demonstrate that the industry can produce good things for communities. In South America communities seem to assume that to some recognizable degree a mining operation will behave like a patron or sponsor. In Australia there has been an emphasis on making agreements with local people, perhaps reflecting the role the judicial system has played in the development and protection of Aboriginal rights to land and protection.

Community relations personnel who are cultural insiders can tell a joke that makes local people laugh. They know when to shake hands and what to say when they meet people, when and how to give praise, how to censure and discourage, how to give instructions, how to make purchases, when to accept and when to refuse invitations, and when to take statements seriously and when not. These cues, which may be words, gestures, facial expressions, customs, or norms, are acquired by all of us in the course of growing up and are as much a part of our culture as the language we speak or the beliefs we accept. All of us depend for peace of mind and our efficiency on hundreds of these cues, most of which we do not think about. When an individual enters a strange culture, all or most of these familiar cues are removed. He or she is like a fish out of water. No matter how broadminded or full of goodwill the individual may be, a series of props has been removed. This is usually followed by a feeling of frustration and anxiety.

Both companies and local people benefit when communities personnel exhibit a sense of service which cannot be assumed as a consequence of employment. It is not produced by a pay rise or a productivity agreement; it is not ordinarily an outcome or reorganization or rationalization or downsizing. A sense of service comes from selecting the right people, those who have sensitivity and feeling for communities and relationships with communities. It also derives from a process of long organizational maturation, one where those involved have over and over again seen good work, and how it is done and bad work discouraged. The most important activity may well be the one where there is nobody to watch, no paper trail, no residue for purposes of accountability: a communities professional official visits a distant town or village, spending time with those who need help; a teacher goes out of her way to help a pupil who is not in her class; an examiner given a script to review changes the marking though no superior will know what she or he has done. Extra effort is not only vital; it is the very substance of the work. Creativity, a broad smile, a helping hand, these are the things that come from within and they are the product of self-discipline and a sense of service.

## Baselines and the Environment

Twenty years ago, as Bougainville showed, community perceptions about the environment often received scant respect from environmental scientists because they were not seen as science. There was no strong attempt to comprehensively understand community environmental ideas or to have local people participate in decisions about their environment.[10] When Diavik Diamond Mine Inc. (DDMI) in Canada was going through the permitting process it was necessary to demonstrate to First Nations groups that an oil pipeline would not interfere with the passage of caribou. Experiments were conducted by the Dene Cultural Institute and funded by the miners which showed that caribou could either jump over or go under the pipeline.

Local perceptions about water, its use, the extent of underground water resources, and rates of replenishment in mining projects in Mongolia, Turkey, and Borneo needed special attention because the environmental science produced information that was often very different from the understanding of local people about the nature of underground water resources. In Mongolia local people feared that mining would exhaust underground water resources; in Turkey local people assumed that the company was taking water out of the earth in the night and transporting it elsewhere.

On Lihir Island in PNG environmentalists paid little attention when the islanders said that mining was raising the sea level and adversely affecting the ability of turtles to come ashore and lay their eggs. Because the company continued to behave in ways that suggested that this claim was not taken seriously local people refused to give permission for activities and engineering that the mine needed to maintain production. Company environmentalists gradually learned that local beliefs about the environment needed to be treated with respect.[11]

## COMMUNITY CONSULTATION

The social baseline was expected to provide an authoritative and accurate explanation of how decisions were made in the community. Customary decision-making had to be described in detail so that one could understand how one could know the will of the people including how both agreement and disagreement were signified. This was made available to all company personnel as well as, when necessary, to government, civil society, and aid agency personnel. The company had to be able to understand and record ways in which decisions were made and to identify the various areas of community life that might be subject to different modes of decision-making.

Westminster-style one-man and one-vote democracy and focus groups might not fit with local cultural norms, and taking advice from those who always tell visitors what they think they want to hear might not be wise. How could a company maintain consultation and negotiation with the community throughout the life of the mine? What structures and institutional processes could be used? How much of the outside world did the community understand? Who was involved with community decision-making? How did the community make decisions and how long did decision-making take? How was agreement and disagreement signified? How much of the outside world did the community understand?

Was what was being asked by outsiders in relation to a project or other development a decision of a sort that the community had never made and therefore there was no precedent to follow? Did they believe all must agree or that there could be minority dissent? Who could decide and who could not? Did individuals have opinions that counted? Did women have opinions that counted? Did young people or old people have views that counted? Was the number of community members who had an opinion important or did social class and authority give some a greater say than others? To what extent was there factionalism? Should the information be supplied in writing and in a local language? Who should be involved and how? Would a verbal presentation be best?

In Rajasthan, India, I watched as aid agency personnel who were investigating rural development held an evening meeting with those they thought were the local opinion leaders. They suggested to the audience that local people needed clean water, primary education, and birth control. The meeting did not disagree, and the mission members left to report to their headquarters. I stayed behind to talk with the women who were clearing up. Why didn't they seem to be pleased with the meeting, I asked? Hadn't they agreed that what would be provided would be helpful? Eventually, they admitted that it did not seem right to disagree with powerful people like those on the aid mission. No, what they had agreed to as their big needs was not really their top priority. Their highest priority was a women's toilet. At night, they had to send their young daughters a long way to the nearest latrine and the girls were often molested on the journey.

Listening is hard work. Those who are good listeners pay attention to what Hugh Hall called "silent language." Silent language refers to the cues and signals from body movement and posture that accompany conversation. Geoffrey Masefield, who worked in communities in Uganda said that "In many tropical societies there are set openings to conversations, rather as in a chess game, the observance of which is regarded as a part of normal good manners. There are conventional phrases of greeting, and conventional answers to them, sometimes to a depth of three or four remarks on either side. Next, it may be customary to ask if rain has fallen locally, and whether peace prevails; one may go on to enquire about the health of family and relatives; not until these preliminaries have been completed, and then often only slowly and obliquely, does one get round to the real purpose of the visit."[12] Failure to observe these niceties, such as by a busy mining executive who wants to get straight to the purpose of his visit, makes the visitor appear uncouth and uncivilized.

Good manners help the consultation process. Julius Nyerere, who was President of post-independence Tanganyika (later Tanzania) and a shrewd observer of the process of contact between officials and the public, said that the way instructions are handed out could make all the difference in the world to the man whom they intimately affect. When a civil servant was dealing with the public directly his courtesy made an essential difference both to his own position, and to the understanding with which his words were received.[13]

## Subanun Community Decisions

A disagreement in the Philippines showed the value of having information about consultation and how remote traditional communities made decisions. On the island of Mindanao and in the Zamboanga province of the Philippines Rio Tinto had an ongoing exploration project on land belonging to people called the Subanun, an upland community like the Montagnyard of Vietnam. A group of Irish missionaries from the Columban order mounted a furious campaign against Rio Tinto in the Philippines and the UK claiming that local people did not want any mining to take place on their land. The Columbans campaigned to get support from UK politicians, labor unions, and Rio Tinto shareholders. They attended Annual General Meetings of shareholders where they repeated loudly and often that the company was not wanted:

> At the May 1997 Rio Tinto AGM in London, a letter from Pagadian's diocesan Bishop Zacharias Jimenez, supporting corporate withdrawal was read out to shareholders. Rio Tinto requested a dialog with the Bishop and, on June 30, a meeting was convened between concerned groups and company representatives. Professor Glynn Cochrane, an anthropologist, who is head of Rio Tinto's community development, led the company delegation. Subanun leaders chose not to attend because of their experiences in being misrepresented by the company. All those present at the meeting—Church, NGO and settler community groups—expressed what Professor Cochrane acknowledged to be "deep felt opposition" to Rio Tinto's plans. The company seemed unable to accept this expression of demand to quit the area: Cochrane's response was that the company board in London would make the final decision.[14]

Exploration personnel who had worked in the area for some years were adamant that the Columbans were not correct in their view about local opinion. To get a third-party opinion Rio Tinto asked Chuck Frake, a Yale anthropologist, to visit Zamboanga to discern whether or not they had made the decision the Columbans claimed that they had made. Frake had worked with Subanun for over 30 years and was the author of a popular undergraduate article "How to order a drink in Subanun."[15] Frake's report described the structure and organization of Subanun society. He pointed out that it was an acephalous society, one without hereditary or other strong leaders. The population of fewer than 200,000 people was composed of a number of dispersed bilateral kindreds. Anyone wishing to say the Subanun want this, or do not want that, would have had to trudge around in the bush for quite a few months visiting and having small meetings in order to gain a representative sample of what it was that Subanun wanted. Obviously the Columbans had not done this, since most of the population were animist or Muslim and fewer than half were Catholic.

In view of the growing threat of violence caused by the Columban campaign Rio Tinto decided to end exploration and to leave the Philippines. After Rio Tinto left small Philippine mining companies, in which it was said national politicians had an interest, wanted to continue exploration. The Columbans then came back to Rio Tinto for assistance because these companies were reputed to have low environmental and safety standards.

## LOCAL DEVELOPMENT

Spending more in the community may not be the answer because increasing expenditures can have perverse effects when there is no trust, or when local people think they are owed jobs by the company or where there is a culture of compensation. With the best will in the world it is perfectly possible to provide jobs, pay lots of compensation, and do very well with social and economic development but to still end up losing local support. Where a community begins a relationship with a mining company feeling that the company has taken its resources in the ground without adequate compensation, or is thought to have adequate mitigation, or is providing jobs and training to outsiders rather than local people then what the company has is a worrying relationship deficit that needs to be recognized and narrowed.

Reciprocity must be considered because it makes no sense for companies to imply that they give out of the goodness of their heart; it makes social and commercial sense for companies to say what they want or expect as a result of helping, providing, or giving. Helping without any expectation of a return does not make social or commercial sense. Nobody in a community believes in giving without any expectation of a return. And goodwill cannot be assumed. For this reason it is a useful discipline for those who are accountable for the assistance given to communities to have to put down on paper the ways in which they expect that the company will benefit in return for this help.

The human constraints on development do need to be carefully examined in order to avoid waste and disappointment. Mining companies assume that when local people are given an opportunity they would make good use of it. But there are instances where communities view their social, economic, and natural universe as one in which all of the desired things in life, for instance land, wealth, health, friendship and love, honor, respect and status, power and influence, security and safety, exist in finite quantity and are always in short supply. Not only do these and all other "good things" exist in finite and limited quantities, in addition there is no way directly within an individual's power to increase the available quantities. It follows that an individual or a family can improve a position only at the expense of others. Hence an apparent relative improvement in someone's position with respect to any "good" is viewed as a threat to the entire community; any significant improvement is perceived, not as a threat to an individual or a family alone but as a threat to all individuals and families.[16]

Mining and the rituals of an industrial society are not always at the center of local lives and local ambition. The sense of time that local people possess as well as their manual skills and dexterity may owe something to traditional agricultural pursuits rather than shift-work and the eight-hour day.[17] In South America an engineer looking at the employability of the local population for a new mine noticed the way a gaucho was stringing wire on a cattle fence and said he could train the man for a mining job. "See the way he manages to get the wire taut, that's not at all easy," he said. Literacy and numeracy are not always a bar to employment. In Namibia a mine turned men who could not read or write into effective workers by redesigning the jobs so that the color of the wires could indicate what had to be done. Culturally neutral assessment systems have been developed to test hearing, hand and eye coordination, ability to judge speed and direction, as well as peripheral vision. When these tests were used in Madagascar local people who had received minimal schooling scored higher than educated workers from South Africa.

Money is factitive and is owned or possessed by individuals, whereas community well-being is indivisible and belongs to all. Money is combined and divided in ways that result in a loss of social cohesion; some individuals receive more than their share and others not enough. Each mining project generates problems that arise from local levels of distribution. Here the chief difficulty is to determine how widely or narrowly the pool of recipients or beneficiaries should be defined. The larger the pool the smaller the shares in the overall package with the risk that some who feel they have the strongest claims may find their shares diluted below a level they believe acceptable.[18] To avoid the problems created when a collective good, like money, is given to individuals it is sometimes useful to try to use the funds to provide a health center, a water supply system, or a generator that can benefit all community members. This has been done by mining companies in Africa and in Indonesia with some success. However, wells and standpipes have often been introduced without understanding user behavior. Wells are easily polluted when the community has no history of well use and disease can easily spread if the introduction of these well-meaning innovations is not accompanied by a lot of hands-on assistance.[19]

The sudden introduction and distribution of large sums of money has often prompted dependency and bad habits. New employees in Africa and Asia unfamiliar with the use of money and credit have run up large debts on white goods and household furniture. In a number of African countries as many as half the employees at operations have had up to half their monthly take-home pay garnished under court orders. It is necessary to pay attention to the demand for money. Economists have been saying for years that the imposition of taxation would encourage people to work to pay those taxes and in so doing they would acquire a taste for money and this would promote economic development. Seasoned administrators know that the demand for money may be limited to the individual's need to buy a bicycle or ammunition for a shotgun and that once those funds are secured the interest in money-making declines. Local people do indeed respond to money-making opportunities with gusto but when they have earned the amount of money they want to earn to buy a battery or a bicycle they stop working—in the words of the economist the elasticity of demand for money is low and as a result well-meaning investment hopes are dashed. In Canada the payment of oil and gas royalties to individuals who were unused to such incomes frequently resulted in excessive use of alcohol and the breakdown of traditional discipline. Among mine workers in Namibia, "Wealth is not gained as an individual but as a star of the team. Workers are forced to identify and demonstrate brotherhood, since non-compliance signals that the individual's loyalties do not lie with his fellow workers but rather with himself or the whites, and that he does not wish for, or need, their social and psychological support."[20]

What is needed is an approach which encourages communities to take responsibility for their own future while providing an opportunity for them to be able to learn their way out of situations they want to change. This is an interesting idea recently illustrated by a population program in Ethiopia that has had extraordinary success by assuming that population control requires the community to understand their own responsibility and power in relation to contraception. Called Learning Our Way Out and run by the International Institute for Rural Reconstruction, from the Philippines, it has successfully trialed this innovative program, which has a philosophy that has potential importance far beyond population control.[21] The community works together to understand the benefits and burdens of an increasing population. Conscious decisions are made to limit family size by family members whose opinions are important. Community members support each other's decision.

What is it that local people want and are prepared to work hard to get? What can they do individually or in groups? What should be the basic organizational structure, a foundation or a cooperative, for example, through which innovation should be routed? Should a new organization be created or will existing local government or village organization be sufficient (community does not necessarily reflect an homogenous entity with common ideals and living in harmony)? The baseline should provide an understanding of the division of labor, specialization, resource distribution, the economic calendar, hunting, gathering and cultivation, and ways of organizing for economic work.

Mining companies need to avoid the lazy use of needs analysis to identify areas for the provision of company assistance. Needs analysis is simply an assessment of what sorts of help and assistance a community needs—health, education, roads, jobs, electricity, for example. Which community does not need improved health, education, or economic opportunity? But so do communities in London, New York, and Sydney. Of course, communities in rich and poor countries need medicines, investment, and education, but the material deprivation of their lives is not the only thing those who wish to help need to know about. It is necessary to know exactly what sorts of health or education is needed and the extent to which groups and communities can help themselves.

Knowing what communities can bring to any improvement attempt is essential if assistance is to succeed. Agricultural knowledge may be strength in one area, manual skills in another. Local people may have high or low amounts of stamina; they may possess good humor, fortitude, discipline, a willingness to work with outsiders, encompassing memory, creativeness, and reasoning power. There is a need to balance assumptions about what communities do not have with assessments of any strength. Appreciations which dwell on what it is that communities do not have to the exclusion of information on what they possess are not enough: we also need to understand community assets.

Logic suggests that when herders and pastoralists make money they will improve their livestock. But this did not happen in the Sahel region in Africa or the Gobi desert in Mongolia, where farmers making more money buy more animals, thereby reducing the overall quality of their herd. They do this because it is the number of animals a man possesses that is prized rather than the quality of his livestock. The older generation, those with the most seniority and status in traditional society, often end up with the least money and status after the introduction of money. The young earning money see their elders standing in the way of change and traditional systems of authority begin to break down.

UN agencies have shown little interest in community development though companies tried to get them involved. Mining companies that did not do their homework on the World Bank Group (WBG) often ended up working with the wrong bit of the WBG. They sometimes began by liaising with the International Finance Corporation (IFC) when a World Bank connection would have made more sense. The IFC did not have contact with Ministries of Finance in developing countries, since that was the World Bank's role and it was the World Bank that had the best country economic analysis. Nor could the IFC help with changing government policy on intergovernmental transfers to ensure that more of the money raised by mining in a region was returned to that region; again, that was the International Monetary Fund's (IMF) role. The IFC makes private sector investments and the World Bank (or, in more formal terms, the International Bank for Reconstruction and Development [IBRD]) makes loans to governments. Developing countries had separate IFC and IBRD representatives, and each had their own projects and priorities.

Both IFC and IBRD wanted to play a key role in Rio Tinto's mega mining investments in Guinea and Mongolia. The IFC, which in the 1990s had quite limited social and economic analysis capacity, wanted mining company business. Senior managers in Rio Tinto were persuaded that getting their Good Housekeeping seal of approval[22] could be very useful. The decision to cooperate with international organizations represented a shift from the 1995 position when big mining companies such as Rio Tinto had wanted little to do with the World Bank or UNDP. The World Bank was thought to be very slow-moving and bureaucratic. The first contacts between Rio Tinto and the World Bank were stiff, reserved affairs because each was used to being top dog.

Rio Tinto was right to be suspicious because when there is talk of a partnership with the WBG it is the bank that is always in charge and where that is not the case there is no partnership. When Rio Tinto wanted to raise something with the World Bank they called the President. MMSD produced a slight thaw, and then came Madagascar, where the World Bank and Rio Tinto cooperated on a mineral sands project. Both parties cooperated because each was able to stretch its dollars. The miners were able to site their infrastructure for the project in a way that helped the company and the World Bank. The production of electricity and clean water also helped the relationship, though the idea that the small amount the World Bank was adding to the investment package entitled them to have their procurement rules followed for all Rio Tinto's project expenditures did not.

## PARTNERSHIPS

In 1998 mining companies were being persuaded by the UN and the Prince of Wales Business Leaders Forum that partnerships were the answer to good community relations. Partnerships did help in some situations where community relations were in place.[23] In addition, a number of excellent institutional arrangements have been developed by Rio Tinto to help with the distribution of money and benefits, and the most popular of these has been the foundation. In Indonesia Yayasans have performed a similar role to a foundation and in Australia there has been interest in using trusts. Since the 1950s, the Rössing Foundation has provided an excellent example of what can be done. It was joined by the Palabora Foundation at the Palabora copper mine in South Africa, Bougainville Copper Limited, the Escondida Foundation at BHP, Rio Tinto's joint operation in Chile, and the foundation at Kelian Equatorial Mine in Indonesia. These foundations have done excellent work and have managed to remain relevant by responding to changes in their operating environment. In some cases these foundations have been in addition to the provision of company community relations though at Phalaborwa the foundation made so much money from its endowment that it was financially self-supporting to the point where there was concern about the survival of its tax-free status as a foundation.[24]

Where a foundation simply funds good ideas and then leaves the execution to others the agreement may work well and be responsive to community welfare. However, when the institution becomes involved with the day-to-day running of health care and the funding and construction of capital projects (which have the capacity to be labor-intensive) the threat of institutional capture by a small local elite and the subversion of the original purpose of the organization may be an ever-present danger.

Very useful non-legal agreements or understandings have been reached by a mining company and a community to contribute to local improvement. Betterment agreements can exceed US$100 million a year as in the case of Freeport's Indonesian mine or the large BHP/Rio Tinto joint venture Escondida mine. These may best be thought of as a partnership or a cooperative agreement. Keeping your word and gaining and keeping trust does not need much formalism nor does the operation of the court of public opinion. For a mining company with a mine that generates enormous wealth the practical and political requirement is to be seen not just to be making an agreement but also successfully expending substantial effort to return part of the wealth generated to local people. To decide what level of resourcing may be justified betterment agreements need to be examined in the light of what stage the operation is at in the project cycle and whether a key event such as an extension or mine closure is imminent. However, in most cases the resources deployed will usually be far fewer than those used to secure an agreement to mine.

## Supply Chain

There are two parts to developing a baseline supply chain. First is to develop the business case for the company that shows the value in a community supply chain. It would be necessary to work closely with a number of departments and management to get a good idea of what is needed, the quantities needed, the anticipated budget if there is no community supply chain, and some rough sense of the figure for local spend. This requires a detailed knowledge of mining, an understanding of whether or not large items and expenditures could be broken down in such a way as to provide local opportunities, a timetable for the development of the overall supply chain to see if at early stages there might be needs that could be met locally, and a sense of where suppliers are located around the world and what they could and could not supply at competitive prices.[25]

What is needed is an assessment of the local skills and capacities, including management capacity, to estimate supply chain potential. This should be completed before beginning to hand out small contracts. An excellent business incubator program in Asia failed to make progress because one or two entrepreneurial workers cornered the market before other potential suppliers had a chance to show what they could do. An "E"-commerce initiative in Southern Africa discovered that only one in ten of the local population could meet the performance requirements, leading its promoters to conclude that additional supply opportunities had to be located before any launch.

Achieving some sort of balance between these two approaches is essential. If the technical demand analysis is superficial then high-value bulk purchasing from elsewhere would be the norm, resulting in only labor-intensive low-value input opportunities for local communities, such as mess hall food supply, hauling and trucking, minor earthmoving, and items such as clothing, survey pegs for geologists, or fencing. While these sorts of contract would inject cash into the community they might do too little to contribute to future supply chain development. To achieve that a five-to-ten-year supply chain plan is needed, starting with an appreciation of what the community would be unlikely to produce, say sulfuric acid or sophisticated instrumentation. However, the plan should also envisage which kinds of items, with assistance and support, the community might be able to supply after a number of years of training.

To succeed with building supply chain opportunities there must be active, visible, and strong top management involvement and incentives. Maintaining focus on the business case helps to manage change throughout the project, to keep people focused, aligned, and moving in the right direction, and to make sure that the expected benefits are achieved. It requires imagination and creativity to persuade managers busy doing other things that they should opt for local solutions that need to be built bit by bit, instead of choosing the integrated and bolt-on solutions with extensive functionality that are available off-the-shelf. These in many cases would eliminate the need for customized solutions which are costly, take a lot of time to develop and implement, and fail to cover the supply chain spectrum. Close coordination is needed between communities personnel and those involved with global purchasing and supply, but most organizational charts do not allow for this kind of contingency.

As big mining companies have moved toward centralized purchasing the difficulties in securing local contracts have grown. In one instance a community was promised that a small company making pumps and another supplying tires would benefit from the company presence. Two weeks later shame-faced community personnel said that all the companies' needs for pumps and tires were now being supplied under a global agreement with an overseas supplier. The lesson is that community personnel interested in procurement need to keep in close and constant touch with their company's global supply team.

## Mining Wealth Goes Elsewhere

Even in industrialized countries it is extremely difficult for a mining company to ensure that a major portion of the wealth from mining can be retained locally. Since it is often the case that much of the materials and services needed by a mine are not produced locally local wealth retention needs to be carefully planned and implemented over a number of years. The Pilbara region of Western Australia where iron ore mining takes place covers 502,000 square kilometers and a decade ago was populated by only 40,000 people. The proportion of the indigenous population increased from 12 percent in 1991 to 15 percent in 2001, largely as a result of the total population declining. Overall prices for goods were 11.3 percent higher than in Perth and median housing prices were far higher than the Western Australian average. As a result of the iron ore boom which had a strong national benefit could the rising economic tide lift all ships with the result that Aboriginal people have become participants in the mainstream economic life of Australia?[26] However, the Pilbara's relatively narrow economic base meant that, overall, the region received relatively little economic benefit from an increase in iron ore demand.

In 2005 the Gross Regional Product (GRP) of the Pilbara was approximately A$12.9 billion, or over 17 percent of the Western Australian economy[27] and of this perhaps less than A$200 million was retained locally. Wealth generated by Pilbara Iron bypassed the Pilbara on its way to Perth. Modern mining is capital-intensive and skill-intensive. Little mineral processing occurred locally and few supply and skill needs were met locally. In 2004, Pilbara Iron's demand for materials, facilities, and services was A$915 million (US$704 million), or 33 percent of total revenue. Nine percent of total purchases were supplied locally; 52 percent came from elsewhere in Western Australia; 37 percent from other parts of Australia; and only 2 percent from overseas.

Pilbara Iron represented a significant part of the Pilbara economy. It employed 13 percent of the workforce (Pilbara Iron is the largest employer of resident workers) and 7 percent of the total Aboriginal workforce. The company also provided 17 percent of the Pilbara GRP and purchases at least A$53 million of production locally. While in absolute terms the overall economic contribution from the sector is high, in relative terms mining multipliers were not. Modern, capital-intensive mining added more value to the economy per unit of labor than did other industries. Conversely, this meant that payments to suppliers for services and materials were relatively low. Payments to suppliers are the most commonly used proxy for backward linkages, which in turn determine the weight of a multiplier.

Input/output analysis showed that the underdeveloped regional industry and service sectors restricted the Pilbara's ability to extract benefits from mineral production. To put it differently, it appeared that there was relatively little scope for diversification—the local economy depended on direct contributions from mining. This experience suggested that the local impact of mining needed to be carefully monitored because, given the capital- and skill-intensive nature of the industry, one could not assume that supply chain and employment would automatically benefit the local economy.

## DIAVIK DIAMOND MINE INC.

Over the course of several years Diavik Mining Company[28] had undertaken a social baseline to build company and government understanding of the Indian groups involved—Dene, Dogrib, and Yellowknives. The notion of sequence was important—realization that social relationships characterized by mutual understanding and respect had to be in place before getting down to the details of any agreement. Consultation with the Chiefs of the Indian groups was lengthy and complex. The Bureau of Indian Affairs (BIA) of the Canadian government was also involved. Canada's approach to oil discovery and the Gwichin Indian Agreement had caused a boom in northern development. As a consequence the Indians pushed to see if the mining company could be prevailed on to up the terms of deals with the government such as treaties 9 and 10.[29]

The DDMI demonstrated that a successful development relationship with communities was not just about jobs and supply chain opportunities; the partnership was comprehensive since it covered not only economic aspects but also welfare, cultural, and social practices. The aim was to develop and preserve the way of life of the Indian groups in ways that they supported.

The First Nations Indians were grateful that exploration teams and company personnel took Traditional Environmental Knowledge[30] seriously and took concrete measures to preserve items of archaeological significance such as caches close to porterage sites on the river system and by DDMI funding of the Dene Cultural Institute. Indians wanted the company to be on their side over caribou hunting when that was threatened. While caribou hunting was claimed as a very traditional activity it could also encompass modern aspects such as shooting caribou from trucks on the ice road.

The mine situated in the North Slave Region of the Northwest Territories, Canada, about 300 kilometers (190 miles) north of Yellowknife, began operating in early 2003 to mine four diamond-bearing kimberlite pipes. The mining infrastructure was constructed on one of two large islands in a large arctic lake, Lac de Gras. Construction material as well as gasoline and diesel had to be brought in on the ice road during the winter.[31] The infrastructure included a 2000-meter airstrip, diamond recovery process plant, processed kimberlite containment area, accommodation blocks, water and sewage treatment facilities, and power generation facilities. The diamond-bearing kimberlite pipes were located just offshore of the island, and access was made possible through the construction of water retention dykes to create artificial impoundments that would provide surface access for mining.

One of the early initiatives was to take training to all the Indian settlements in order to assess aptitude and then, on the basis of testing, to provide training in the settlement so that when a job became available there would be no delays. In making payments and compensation the company had to be mindful of the debilitating effects of alcoholism and gambling evidenced by the impacts of oil and gas royalty payments in Ontario and elsewhere on Indian welfare. Indian leaders were taken to see the problems and solutions worked out by Indians themselves such as "Spirit Lodges" and "healing circles" in big Canadian cities at first hand. Money was needed to maintain group solidarity rather than be frittered away on individual claimants with nothing to do but watch TV and drink.

The Community Team of the company repeatedly visited villages and kept talking. Small things helped as when a chief's son was given tickets to an important ice hockey game or going fishing in the summer. Agreement was reached in 1999 with Tlicho, Yellowknives, Dene First Nation, the North Slave Metis Alliance, Kitikmeot Inuit Association, and the Lutsel K'e Dene First Nation. The company made a commitment to local training, employment, and business benefits for Northwest Territories and the West Kitikmeot region of Nunavut communities. Participation agreements were really understandings rather than formal legal agreements. Initially, the Indian groups were keen to have one of the Benefit Impact Agreements that had become popular in Canada. DDMI's concern was that the commercial aspects of the project were not certain. What if the Indians settled for a figure that in the light of production turned out to be too low? If the relationship was simply legal the company would not make a change. But if there was a good relationship then the agreement could be mutually adjusted.

It was a testament to the quality of the relationship that participation agreements, which were about intentions and principles, were executed with each of the Indian groups. In later years these agreements were adjusted upwards when production exceeded targets. During construction laundry had to be sent to Edmonton as there were so few local services. However, by the end of the first decade of operation C$1.9 billion was paid to local suppliers.

A review conducted by the Oxford consulting company Synergy[32] and the International Business Leaders Forum review concluded in 2003 that while the proof of the success of their participatory inclusive approach to project development and commitment to environmental excellence will only be proven over time, the initial signs were encouraging.

## A RELATIONSHIP PLATFORM

As soon as it is clear that a mining project will proceed, the advantages derived from building of relationships that are characterized by mutual trust and understanding is not just in the fact that this relationship building will create a good climate for discussion with local people over the development of a mine but also because such work can provide a platform from which a number of initiatives can eventually be launched. These may range from mine extension or closure to the development of public health programs or micro-enterprise initiatives. A relationship platform is created by building relationships with a broad spectrum of individuals in the community—leaders and opinion makers, high- and low-status individuals, women, and men.

In the vicinity of Oxiana Resources' Sepon exploration camp near the border with Vietnam, and close to the Ho Chi Minh trail, the Oxiana Resources Mining Company had to build good relationships with local people in order to persuade them not to practice "slash and burn" horticulture approaches. The company needed to discourage slash and burn, not because of any prejudice against this as a form of agriculture but because of the danger. The entire region had been affected by the Vietnam War. Lao Peoples Democratic Republic (PDR) is the most heavily bombed country per capita in the world. An estimated 260 million bombs were dropped on it between 1964 and 1973. Of these, some 80 million did not explode, leaving a deadly unexploded ordinance or UXO legacy. UXO continues to claim lives and cause injuries in the Lao PDR. In the vicinity of Sepon's operation there was a great deal of unexploded ordinance. US bombers returning to their bases from bombing raids over Vietnam often jettisoned bombs they had not used. Many of these bombs were small and had plastic casings. They were extremely hard to detect but, when "slash and burn" horticulture was employed, they exploded and many people including young children lost limbs.

In Madagascar good relationships with local people have been essential to the company's task of persuading village people not to cut down hardwood trees to make charcoal. Here a Rio Tinto ilmenite operation used satellite photography that showed a small remaining coastal strip was vanishing at such a rate that within a few years nothing would have been left of the island's rare timber. Day by day felled timber was carried from the woods on the back of village people to the towns where it was turned into charcoal. The city of Fort Dauphin consumed 10,000 tons each year.[33]

The different types of firewood-producing trees that were offered by aid agencies in the hope of substituting new timbers for old had little appeal for local people, since lots of smoke was produced and it took longer to get sufficient heat to cook. The French colonialists' answer to the tree problem on Madagascar had been to create forest reserves or preserves that were off-limits to village people. Those who entered the forests to cut timber were prosecuted. Rio Tinto believed that making forests inaccessible to local people would not work. As an alternative to interdiction the mining company established a nursery and supplied hardwood seedlings to village people who became major participants in the replanting program. Woodlots that could provide more acceptable cooking material were provided. The key to success was putting the village people in charge of the replanted areas and the forest woodlots. This was done by concluding a traditional social contract called a Dina with the village people. The Dina has been respected and the village people have become part of the solution rather than part of the problem.

Obviously having a relationship platform in place will be of great assistance when negotiating mine start-up and prior to closure. It can also be helpful during the life of the mine for extensions and expansions. One company in the USA that had built a relationship platform over many years avoided a US Environmental Protection Agency (EPA) Superfund listing which would have caused great inconvenience and expense because the community made representations on their behalf. In Namibia relationships with the Rössing Foundation had been building since its establishment in 1978 provided a platform from which new initiatives could be developed. David Godfrey OBE,[34] Director of the Rössing Foundation for many years, explained how the platform was built: "It was the small things we did that mattered the most." From this platform built out of doing small things in hundreds of relationships the Luderitz initiative was launched to develop seamanship and the Environmental Education program provided the basis from which the Community-based Natural

Resource Management program later developed and integrated rural development in the Erongo area. A Master Maths program was launched at the behest of the government of Namibia.

## NOTES

1. Useful material on quantitative and qualitative approaches can be found in *The Manager's Guide to Data Collection*: prepared for the Agency for International Development and looking at community-level data required for poverty alleviation authored by Hageboeck et al. (1979).
2. See Carstairs (1958).
3. The Summer Institute of Linguistics (SIL) which has done a great deal of work on the translation of the Bible into little-known languages accepts that change in traditional societies is inevitable and argues that there is a role for organizations who wish to moderate and influence the speed, direction, and pace of that change so as to minimize discomfort.
4. When I was working in the Santa Cruz islands in the Solomons I noticed that anthropological text books frequently mentioned "red feather money" even through red feather money had not been used for almost 100 years.
5. My friend Dr. Angus Green, a careful and highly experienced observer of Indonesian and Malaysian society, together with a number of other Malaysian anthropologists, provided an excellent social baseline. Angus became a formidable advocate for those wanting to see the government do the right thing by the Penan.
6. The *Contemporary Ethnography* was made available to the government of Sarawak, NGOs, and interested researchers. Parts of the work were published by the members of the research team.
7. See Goethe (1867).
8. See Hicks (1964), Cochrane (1969), Lunenburg (2010).
9. See Erasmus (1961).
10. See Kirsch's (2014) critical account of the mining industry. The innovative approach of Brokensha et al. (1980), see Cochrane's review in *American Anthropologist* Vol. 84, Issue 2, in 1982, or the seminal earlier work by Forde (1934).
11. See Skalnik (1989).
12. See Masefield (1976).
13. Appendix 2, Presidential Circular SHC/C.230/8, State House Dar-es-Salaam, 21 October, 1963.
14. See International Federation of Chemical Engineering and Mine Workers (1997).
15. See Frake (1964).
16. See Foster (1965).

17. While teaching in the USA I noticed that an academic colleague, who was blind and black, was very tough on black students when they were late for his class. I asked why and he told me that the parents of these students had been waiters and maids. Those jobs had required strict punctuality, a trait he was sure they would have passed on to their children.
18. See Jorgenson (1997).
19. See Spicer (1952).
20. See Gordon (1978).
21. Personal communication from Jane Boorstein of Columbia University who is a Director of the International Institute for rural Reconstruction and the designer of the Ethiopian population project funded by the Gates Foundation.
22. IFC did not have contact with Ministries of Finance in developing countries since that was the World Bank's role. Nor could the IFC help with changing government policy on intergovernmental transfers to ensure that more of the money raised by mining in a region went back to that region— again that was the IMF's role.
23. See Tennyson and Wilde (1998), Nelson and Zadek (2000).
24. Ramanie Kunanayagam authored a paper, *Foundations Guidance*, while she was at Rio Tinto.
25. Critics of mining companies have yet to develop the skills to present a detailed criticism of the way supply chain opportunities for communities have been handled. Supply chain presents a much bigger target than the usual candidates for their arithmetic of negligence.
26. In 2004–2005 Rio Tinto commissioned think pieces about the Pilbara economy from the consulting firms Acil Tasman, URS, Curtin University, and the Australian National University. The data was discussed in a number of workshops with academics and the government of Western Australia.
27. A$12.9 billion was the region's GRP estimated from the input-output tables.
28. Led by Doug Willy, who orchestrated contact between the company and the Indian groups and mine manager Rod Davey.
29. See Wright and White (2012).
30. See Inglis (1993).
31. The ice road is formed in winter when the lakes freeze to a depth sufficient to carry heavy vehicles.
32. Synergy had worked closely with the company in a number of locations in Asia and Africa and was unusual in terms of its understanding of the new communities approach and the social competence of its staff.
33. Quebec Iron and Titanium, Madagascar Minerals, Quebec, December 2000.
34. David Godfrey, OBE, was succeeded as Director of the Rössing Foundation by Len Le Roux and later by Tjiho and Grobler (n.d.).

# BIBLIOGRAPHY

Brokensha, David, Michael Warren, and Oswald Werner. 1980. *Indigenous Knowledge Systems and Development*. Lanham, MD: University Press of America.

Carstairs Morris, G. 1958. *The Twice Born: A Study of a Community of High-Caste Hindus*. Bloomington, IN: Indiana University Press.

Cochrane, Glynn. 1969. Strategy in Community Development. *Journal of Developing Areas* 8: 5–12.

Erasmus, Charles. 1961. *Man Takes Control*. Minneapolis: University of Minneapolis Press.

Forde, Daryll. 1934. *Habitat, Economy and Society*. London: Methuen.

Foster, George M. 1965. Peasant Society and the Image of Limited Good. *American Anthropologist* 67(2): 293–315.

Frake, Charles O. 1964. How to Ask for a Drink in Subanun. *American Anthropologist* 66(6), Pt 2: 127–130.

Goethe. 1867. *Faust*, Part 1: 1112, English translation by John Wynniatt Grant, London.

Gordon, Robert. 1978. The Celebration of Ethnicity: A Tribal Fight in a Namibian Mine Compound. In *Ethnicity in Modern Africa*, ed. Brian du Toit. Boulder, CO: Westview Special Studies on Africa.

Hageboeck, Molly, Glynn Cochrane, Lawrence Cooley, and Gerald Hursh-Cēsar. 1979. *The Manager's Guide to Data Collection*. Washington, DC: United States Agency for International Development.

Hicks, Ursula. 1964. *Development from Below*. Oxford: The Clarendon Press.

Inglis, Julian. 1993. *Traditional Ecological Knowledge: Concepts and Cases International Program on Traditional Ecological Knowledge*. Ottawa, Canada: International Development Research Center.

International Federation of Chemical Engineering and Mine Workers. (1997). *Rio Tinto: Tainted Titan, The Stakeholders Report*. Brussels: International Federation of Chemical Engineers and Mine Workers, 28, 29.

Jorgenson, Dan. 1997. Who and What is a Landowner? Mythology and Marking the Ground in a Papua New Guinea Mining Project. *Anthropological Forum* 7(4): 599–627.

Kirsch, Stuart. 2014. *Mining Capitalism: The Relationship between Corporations and Their Critics*. Berkeley, CA: University of California Press.

Lunenburg, Fred C. 2010. Managing Change: The Role of the Change Agent. *International Journal of Management, Business and Administration* 13(1): 1–16.

Masefield, Geoffrey. 1976. Agricultural Extension. In *What We Can Do for Each Other: An Interdisciplinary Approach to Development Anthropology*, ed. Glynn Cochrane. Amsterdam: B. R. Gruner.

Nelson, Jane, and Simon Zadek. 2000. *Partnership Alchemy*. Copenhagen Centre: Copenhagen.

Skalnik, Peter. 1989. Lihir Society on the Eve of Mining Operations: A Long Term Project for Urgent Anthropological Research in Papua New Guinea. *Bulletin of the International Committee on Urgent Anthropological Research,* Nos 32–33. Vienna: UNESCO.

Spicer, Edward H. 1952. *Human Problems in Technological Change: A Casebook.* New York: Wiley.

Tennyson, Ros, and Luke Wilde. 1998. *The Guiding Hand: Brokering Partnerships for Sustainable Development.* London: United Nations Staff College, Prince of Wales Business Leaders Forum.

Tjiho, Job, and Grobler, H. n.d. *Draft History of the Rössing Foundation.* Unpublished Document, Rössing Mine, Swakopmund, Namibia.

Wright, Laura, and Jerry P. White. 2012. Developing Oil and Gas Resources on or near Indigenous Lands in Canada: An Overview of Laws, Treaties, Regulations and Agreements. *The International Indigenous Policy Journal* 3(2), Article 5.

# CHAPTER 7

# Artisanal Mining and Closure

In 1987, Rio Paracatu Mineração (RPM), a subsidiary company of Rio Tinto London, was authorized by Brazil's National Department of Mineral Production (DNPM) and the government of the State of Minas Gerais to begin the industrial exploitation of gold in the region. The Morro do Ouro or mountain of gold open-pit mine at Paracatu was then operated by RPM until Rio Tinto sold the property to a Canadian miner in 2004.

In 1997 Rio Tinto's gold mine at Paracatu was invaded by thousands of illegal miners called *garimpeiro* who were trying to pan for gold in the tailings discharged by the mine for the small amounts of gold that had not been captured by processing.[1] After the tailings are discharged from the mine small-scale artisanal miners try to recover the gold that has not been captured by ore processing by putting the tailings in a pan and mixing them with water. As the liquid is swirled around in the pan, the gold particles, which are heavier than the other materials in the tailings, sink to the bottom of the pan and can be collected as a result of this labor-intensive process.

The *garimpeiro*, who were armed, broke into buildings, stole equipment, and assaulted mine employees. *Garimpeiro* who threatened to shoot employees told security personnel that they would rape their wives and daughters while the men were at work in the mine. A security guard who had smuggled an illegal firearm into the mine shot and killed one intruder. Security personnel had been given training in the use of non-lethal methods of restraint and this illegal use of a firearm was not supposed to happen. This incident resulted in the company being headlined

© The Author(s) 2017

G. Cochrane, *Anthropology in the Mining Industry,*
DOI 10.1007/978-3-319-50310-3_7

on the Four Corners TV program in Australia although the guard had been dismissed and the incident passed to the police for investigation. The Federal Police using shotguns blew the arm off one garimpeiro as he was riding his mule away from the deposition area, laden with tailings.

The Morro do Ouro mine where mining has taken place since 1722 is one of the largest gold mines in Brazil, and also in the world, with proven gold reserves of just under 20 million ounces. Between 1982 and 2000 the mine was producing 180,000 ounces of gold annually. In 1999 the mine produced 50 tons of gold or just under 9 percent of Brazil's gold production. Situated less than 3 kilometers from Paracatu, a city with 80,000 residents, the mine is located in the state of Minas Gerais, 230 kilometers from the capital, Brasilia. Until 1980 the Paracatu economy was based on small farming and artisanal exploitation of the gold deposit existing in the municipality. This small-scale economy that took place in the rivers and streams near Paracatu was the main source of livelihood for several hundred artisanal miners called Quilombola who lived in temporary squatter settlements on the edge of the property. Quilombola were descendants of Afro-Brazilian slaves who escaped from slave plantations that existed in Brazil until the abolition of slavery in 1888.[2]

To the Quilombola, panning in the tailings produced by RPM's industrial mining for gold represented a much easier and more attractive way of earning a living than working in the small streams and rivers. By 1998 the mine was suffering around 11,000 illegal incursions a month with many of the entries taking place at night. Fences and banks protecting the tailings deposition area were broken down, which prompted fears that there could be a catastrophic release of the tailings that had been impounded by the operation. Mercury was being used and this was capable of damaging the health of the artisanal miners as well as the environment.

Mine managers thought that the surge in *garimpo* activity was driven by the high gold price and new technologies. Thus, *garimpo* operations became more intensive, as well as more extensive. In particular, pumps were introduced, as were various types of crushers. The technical improvements secured by the *garimpeiro* were all directed at mining more ore; the gold-recovery processes were still limited to gravity concentration and amalgamation. However, the environmental effects of this modest mechanization were severe, with deforestation, siltation, and land sterilization occurring over wide areas. The *garimpeiro* mining activities developed in a labor-intensive and highly polluting framework, with technologies that demanded little capital. Because the *garimpos* were small- and medium-scale undertakings, they could expand by using mechanical and semi-mechanical technologies for rapid and easy extraction of gold, with little or no attention to the efficiency of the operation or to environmental management. Increased gold output means increased use of mercury. These problems were intensified by the fact that these workers were seldom qualified, in contrast to the highly qualified people employed in formal gold mining. The legal controls were inadequate and the environmental regulations for *garimpeiro* were the same as those for large companies. The entities responsible for application of the regulations and safeguards were poorly equipped for their task. Adoption of more sustainable practices was impeded by the growth in the numbers of the unemployed and the importance of the *garimpeiros* at election time.[3]

RPM's mine management tried to increase the gold-recovery rate in order to make panning less attractive by upgrading their ore processing machinery in the plant. They re-engineered the discharge of tailings to ensure that much of the flow was hidden from easy access underground. Dogs were introduced into the fenced-off tailings deposition area. The mine even had its own panners working inside the mine processing plant to extract as much as they could before the release of the tailings. The gold-recovery rate was boosted into the high 90 percent range by these measures. However, in spite of these efforts there were still far too many incursions. London asked RPM to consider engaging with the *garimpeiro* because avoiding or ignoring artisanal miners was obviously not going to improve the situation.

Obviously, for reasons of health and safety, environmental protection, legality, and the loss of tax revenues, most governments would like to regulate artisanal small-scale mining and mining companies would certainly agree. But, although there is broad agreement on what should be done, there has been much less progress on the issue of how these policy objectives can be secured. The issue of whether or not small-scale mining should be legalized does not seem in the end to make much difference in situations where the small miners are determined to go their own way regardless of whether the activity is seen as legal or illegal.[4] It may make little sense to work out what ought to happen to artisanal mining unless at the same time it is clear that there are boots on the ground as well as trained and committed communities personnel and people who can take suggestions forward. Somebody must work with those to be helped and that somebody must be prepared to alter and change plans and intentions in the light of what it is that these relationships reveal.

RPM had trained community personnel who could work with the *garimpeiro*. Trust had to be built and the perception that the company did not care about the poor people in the town had to be robustly addressed. Face-to-face relationships had to be established between mine personnel and the poorest in the town. Avoidance was not helping because the situation was not self-correcting. Mining companies engaged with their critics, so why not *garimpeiro*?

Neither in their social baseline nor in their five-year plan had RPM covered the *garimpeiro*. To do so a great deal more would have to be known about the *garimpeiro*: who were they, where did they come from, and why did people in the town support them? Since community relations personnel might be regarded with suspicion by the squatters in the artisanal settlements it was decided that a social baseline for the *garimpeiro* should be entrusted to a university. Professor Parry Scott and a small team of anthropologists from the Federal University of Pernambuco were engaged to do the work.[5] The team joined the *garimpeiro* as they were prospecting, lived with them in their settlements and, over a period of months, began to provide mine managers with better understanding of the way the *garimpeiro* worked.

What the anthropologists showed was that well-off mine employees, including the mine manager, seemed to communicate best with well-off townspeople. Those who were well-off in the town had little contact with the poorest people in their town and senior mine personnel had no clear idea of what the company's community relations was trying to do about poverty. This participant observation revealed that the good work being done by the company in the town was not well understood by either Paracatu employees or the poorest people in the town.

Many of the *garimpeiro* turned out to be residents of the town rather than Quilombola from squatter settlements. This accounted for the increased numbers because poor people in the town resorted to artisanal mining when times were hard. The townspeople who became *garimpeiro* also saw themselves as victims of the company. Poor people said the company was so uncaring that they would not even let the *garimpeiro* take away tailings, which they treated as if they were rubbish. The *garimpeiro* had their own negative images of the company. They suspected that many of the company employees who had big houses and new cars had not been honest. Becoming a *garimpeiro* was not like engaging in criminal activities because prospecting was seen in Robin Hood terms as an activity which targeted the rich miners to help the poor in the town. In the masculine world of the town, not having a job was a real blow to manhood and how could a poor unemployed man establish his worth in the eyes of his family and friends?

The benefits from mining were not reaching the poorest. Mining projects generally have a high profile in the economic landscape of a region because of the obvious direct and indirect economic benefits that ensure jobs, contracts, market for supplies, and other resultant multiplier effects. Analysis asked how the beneficial economic impact of an operation on its neighbors could be increased. The focus was on increasing the economic development potential emerging from mining inputs into production and other direct contributions, that is, the purchase of supplies and services, employment and labor remuneration, tax payments, and so on.

Despite the impressive per capita growth brought on by the mine (Paracatu grew at an annual rate of 14 percent during the five years following the mine's commissioning), poverty and social exclusion increased with the result that 55 percent of the population in the town were below the poverty line. Quilombola and other artisanal miners had been affected by new environmental standards which had done away with dredging and riverbank prospecting before the arrival of the company. Then the growth of massive capital-intensive agribusinesses in the region had further reduced rural employment opportunities. The rate of population increase was high so that as time passed the number of unemployed could be expected to continue to rise faster than the number of jobs that could be created.

While RPM's contributions to the economy were substantial, the local economy seemed to lack the conditions to promote business creation and thus unemployment reduction. For instance, with an urban population of 65,000 people and seven banks in town, access to credit was highly restricted (e.g. Banco do Brasil, Paracatu had only 390 clients eligible for business development cash); there was also weak institutional support, and so on. Businesses in the town were not pulling together. In the past there had been a large number of poorly coordinated initiatives, including credit unions, a chamber of commerce, and training institutes. Both state and federal government agencies and institutions seemed remote and uninvolved.

Following the baseline work by Parry Scott and his team it became possible to redesign the reach and function of the communities personnel in order to address the poverty issues of the artisanal miners. Miners and the poor had to get to know and begin to work with each other. RPM started an ecological park and created biodiversity reserves to show its commitment to the environment. To help local people better understand RPM visits were made to schools and children visited the operation. Mine managers began to visit the town on a regular basis to make it clear that they needed help not only to deal with the *garimpeiro* problem but also for the successful conduct of the business. Paracatu joined with other businesses and institutions interested in Paracatu's improvement: the coalition created a development agency the aim of which was to further economic development, promote employment, as well as better communication and understanding of what the mine was trying to do to help local people.

While RPM had a competitive advantage in mining-related activities this was not the case in other areas where jobs and investment were required. The company provided training in metalwork, masonry, electrical, and plumbing skills. RPM had the ability to start small but important projects and then turn them over to others to manage and replicate. The social baseline suggested targeting key problem areas such as unemployed youth in the towns, and the shortage of "know-how" for growing vegetables where what the company could contribute would help to overcome a bottleneck. Then, if the project was technically and financially sound, arrangements could be made to diffuse the results to a wider population. Training was provided in electronics and home electricity, maintenance work, garbage recycling, market gardening, and clothing manufacture. Each of the projects was approved by the community and given targets to be achieved within a one- or two-year period. For example, the market gardening began by trying to produce 10 tons of vegetables. Within a year it was able to produce 200 tons. Free medicines were distributed, and mobile clinics provided free testing for heart and respiratory problems. Employee understanding and support for community investment increased significantly.

The mine did not necessarily have to eliminate poverty; what it had to do was show that it cared about the elimination of poverty. After community relations had been extended to the garimpeiro surveys showed that almost 70 percent of respondents believed that the company's performance in the town was getting better. *Garimpeiro* numbers were reduced from 7129 in 2001 to 11 in 2003 and the number of incursions was reduced from 659 in 2001 to 22 in 2003.

## GRASBERG

The World Bank estimated that Grasberg had the largest number of illegal miners in the world. A thousand panners in the river could generate two or three times that number in support activities and there was general agreement that between 30,000 and 40,000 were involved directly as well as in supply and support. The headcounts took into account the numbers when tailings flow-rates were high, and panning was more profitable, as well as when river levels were low. The hours of work were confined to daylight and this was unusual because elsewhere in Indonesia, such as Rio Tinto's Kelian mine, the panners had small electricity-generating plants capable of supporting 24-hour prospecting. A few of the panners began by living in temporary structures with tarpaulin roofs, which were subsequently converted into permanent residences. Others lived in Timika which, as a result of the mine and the illegal panning, had grown from a few hundred residents in 1967 to over 400,000 and was estimated to be growing at 17 percent a year. As many as 1200 economic migrants arrived each month from elsewhere in Indonesia. Obviously illegal panning was a major contributor to the economy of the town, and this was evidenced by the fact that the gold recovered from panning kept 22 gold shops in business.

PT Freeport Indonesia (PTFI), in which Rio Tinto has a financial interest, operates the Grasberg mine which is largest gold mine and the second-largest copper mine in the world. Grasberg has been producing 61,000 tons of copper and 20 tons of gold a year. The mine is one of Indonesia's largest taxpayers. The company began constructing its mine in Papua in 1967.[6] Despite staggering challenges, PTFI constructed a 74-mile road across a swampy coastal plain rising into steep mountains and built the mine and processing facilities at elevations exceeding 4000 meters. Papua is Indonesia's easternmost province and occupies the western half of New Guinea—the world's largest tropical island.

The mine discharges between 250 and 300,000 tons of tailings into rivers in the highlands each day. The number of panners in the river had grown rapidly because, for a decade or more, the mill recovery rate for gold had been less than 90 percent. The remaining grade in the tailings was low, however, and with current milling rates there was still a reasonable amount of total gold, although this was uneconomic to Freeport at current operating costs. Studies suggested that illegal panning probably recovered 15–20 percent of the gold contained in tailings. Consequently, the financial rewards from panning have been very attractive for individuals as well as the security forces. An agricultural laborer could earn roughly five times more money than he would make elsewhere in Indonesia. A few of the panners earned as much as 6 million Rupiah (US$600 a month). The security forces who were involved, army and police, could collectively earn up to 50–60 percent of their salaries from panning in the river which in total is thought to generate between US$80 and US$100 million annually. The security forces bought gold, supplied rations, mobile phones, and banking services to the panners. A mile-long stretch of tailings had 60–80 kiosks, protected by the security forces, selling food, cigarettes, phone cards, and alcohol.

The technology used by the artisanal miners was relatively simple: woks, wash boards, and some sluicing. There were a few generators that could power air hoses. Mine managers were worried that any large drop in the amount of tailings going into the river might prompt either the use of more sophisticated technology on the part of the panners or the widespread use of mercury. Mercury "sniffers" that could detect the use of mercury were installed in key sections of the river. The use of nitrous oxide was also monitored. Geologists were worried that if the numbers in the river kept increasing then the panners would begin to excavate the gravels at the bottom of the river where a considerable amount of gold was thought to be trapped.

There was increasing evidence that panning posed a danger to the infrastructure and the environment. Panners invaded the processing plant in 2006 and in September 2010 three hooded and armed military personnel invaded the mill and made off with concentrate at gunpoint. Gravel was taken from the levees thereby weakening the banks and exposing the pipes carrying concentrate from the mill to the port and fuel from the lowlands to the mine. These pipes were highly pressurized and could blind or injure anyone cutting into them. However, as the panners gained more experience they were apparently able to cut into these pipes without injury.

The human cost was also considerable. About 15–20 panners were drowned in the river each year despite warnings to keep away during floods and 25 panners were drowned in a single incident when their huts were swept into the river during a heavy downpour. Freeport was expected to help with the funerals of the deceased. Five to 10 men were killed each year in inter-tribal fighting and again Freeport was expected to help with the funerals. Freeport faced criticism because the children of panners did not get an education, instead spending their day in the river helping their parents. Food was stolen by the military from the mess halls in order to feed the panners. Freeport did not provide medical services directly to the panners on the river though they gave grants to hospitals and clinics that did treat panners. Fifteen percent of new referrals to hospitals and clinics in the area were panners who were thought to have been responsible for a 20 percent rise in health-care costs year on year. Panners had 160 meningitis cases in a year—fortunately no fatalities—owing to the crowded sleeping and living conditions. The incidence of malaria, HIV/AIDS, and STDs was much higher among panners than elsewhere on the mine property. Dysentery and typhoid were a risk in such situations. It was also likely that panners would develop lesions as a result of spending long periods standing in the river.

PTFI avoided all contact with the panners because their activities were illegal and also because they did not want to encourage more economic migrants. Yet it was clear that the government was not going to do anything about the panners. Attempts were made to increase the mill recovery rate and to re-engineer the way tailings were discharged so as to reduce the ability of the panners to recover gold at the top of the river where the rewards were known to be high. While these measures helped a little the amount of gold going into the river remained very high.

## PTFI COMMUNITY RELATIONS

PTFI's corporate headquarters in New Orleans was pressured by the increasing number of panners to consider that the company might have to engage with the panners in order to try to reduce costs and damage caused by their activities. The communities team that was in place had the knowledge and the skills to engage with the panners, particularly those who were Papuan. Social baseline material had been collected over a number of years by a number of anthropologists: John Ellenberger on linguistics, Caroline Cook on land tenure and ethno-botany, Tod Harple on the semi-nomadic Kamoro and Kal Mueller on ceremonial life. Freeport had a history of strong hands-on community relations. PTFI's community practitioners were hands-on professionals who worked well with Papuan people and were passionate about their development.[7] The communities team was divided into two sections: Community Liaison Officers (CLOs) and Social and Local Development (SLD). CLOs went to the villages to talk and consult. CLOs did not handle money; they were expected to work on the relationship between PTFI and the local people to ensure that it improved and that there were no nasty surprises. SLD, the second part of the communities team, dealt with site and local development. They invested in small projects and various micro-enterprise initiatives. The company's malaria eradication program was world-class, and its education and scholarship and the technical training provided by the Namangkawi facility was second to none in the mining industry. Progress had been made with Papuan employment. In 1996 there were 640 Papuan employees. This number rose to 1523 by 2000 with an increase in staff numbers from 48 to 114. Today the figure stands at 6281 or 30 percent of the total;

it also includes 420 Papuans employed by contractors as well as 1156 Papuans in Freeport's privatized enterprises.

Mine managers were concerned that the panners might claim title to their homes and gardens if they remained in place for several years. At minimum there was a danger that if they were removed from their panning areas in the river PTFI might be obliged to follow International Resettlement Standards. Freeport had, with great difficulty, managed to acquire local title to land needed for mining. A series of Hak Ulayat land releases took place between 1974 and 1994. In 2000, with the help of local leaders Tom Beanal and Angus Angiabbak, Freeport negotiated an additional "recognisi," which meant that the key landowners, the Amungme, were reconciled to the operation. A Land Trust Fund, eventually known as LEMASA for the Amungme and LEMASCO for the Kamoro, was to be given around US$500,000 a year to be distributed to landowners. This had not been easy because the Indonesian government, which was highly unpopular with Papuans, did not want Freeport to own or compensate for local land.

The Ertsberg, which was claimed by the Amungme as their sacred mountain, was regarded as state land. The extent to which these lands had been occupied by the Amungme since time immemorial was not clear since there was some anthropological opinion suggesting that land in the area might have been only occupied for a few generations rather than the 10 or 12 generations of residence normally taken to convey ownership in Melanesia. If PTFI extended its community relations to the panners, would the company be expected to provide them with a share of the funds provided annually to owners of the land where mining operations took place? The Freeport Fund for Irian Jayan Development (FFIJD) was established in 1996 to pass on wealth generated by the mine to those local tribes who claimed to be the owners of the Grasberg mine. Since the Fund was established, over US$500 million has been given to local people as a result of a PTFI decision to give 1 percent of gross revenues annually to local people. Seven tribes or sukus were recognized as recipients of these funds: the Amungme, Kamoro, Damal, Ekari, Dani, Moni, and Nduga. About 30,000 beneficiaries were involved with the Amungme making up just over 50 percent the number.

## A Social Baseline for Panning

In order to reach a deeper level of understanding about the panners and the likelihood that they could be included in PTFI's community relations a social baseline was commissioned. A team of anthropologists from

Jakarta's Atma Jaya Catholic University was engaged to survey the ethnic origins of workers, incomes and expenditures, patterns of leadership, organization of the work, conflicts, relationships with stakeholders, and health and general well-being. The study covered the upper stretches of the road carrying the tailings, including the town of Tembaggapura, and settlements at Utekini Lama, Berengamme, Lonsoran, Batu Basar, Kembeliu, and Waa Banti. Also included were the Lowlands between Mile 21 and 43 and the Highlands. Below Mile 34 there were mostly groups from elsewhere in Indonesia such as Torajans, Bougis, Floridians, and Kei.

The study confirmed that activity in the river, from the mine to the sea, was controlled by the Dani tribe, or Lanis as they are sometimes known.[8] The Danis used their outstanding organizational capacity to put a functioning river management system for panning in place. The river was divided into sections each with its own section leader. The Dani imposed and collected entry fees as well as fees for the amount of gold collected and sold. They imposed discipline, expelled non-performers, regulated entry, collected fees, and sorted out disputes. These section leaders were recognized by the security forces. With the Danis in control of the river they began to occupy a prominent role in the local political arrangements and maneuvers. The Danis might have occasional and temporary quarrels with other Papuan tribes such as the Amungme but they would also be prepared to join with other Papuans in order to exert leverage on PTFI.

A significant number of personnel in the security forces owed their living to panning. The panners would have liked Freeport to take over the marketing of the gold but this was not something that the security forces were anxious to see happen. While the security forces were useful in terms of controlling inter-tribal fighting they did little to monitor environmental damage or to control the entry of newcomers. The security forces supplied food, building materials for construction of huts, and marketing services. Women played a valuable role in the informal marketing sector as operators of kiosks and in the buying and selling of gold.[9]

PTFI did not include the panners in their community relations programs and activities. Panning within the Contract of Work (COW) area was illegal under Indonesian law and it was thought to be up to the government to effect removal. Yet that did not seem to be about to happen because public opinion in Indonesia was shifting away from the rights of big foreign companies toward doing more to promote and protect small-scale mining. The local government began to appreciate that it had thousands of potential supporters in the river and began a voter registration program.

Although it would have made sense to use the communities team to work with the panners this was not a suggestion that received an enthusiastic response. On several occasions mine management had the communities team assist with the removal of a few panners who were working in dangerous stretches of the river, and they were also involved in preparing large numbers of panners who were to be sent to their homes elsewhere in Papua. The relationship between PTFI security personnel and CLOs was not helped when security personnel worked side by side with the communities people. The security people always wanted to know what was happening in the villages but the community personnel were only too well aware of the fact that if they were seen to be too close to the security forces then their ability to work at close quarters in the village would be compromised.

## Papuan Separatists

Any proposal to engage with the planners had to take the security situation into account. Operasi Papua Merdeka (OPM) separatists who wanted independence for Papua from Indonesia were thought to be supported by the proceeds from panning. In 2005 five Cambridge undergraduates were kidnapped by separatists to draw attention to their demands for independence. The province of Irian Barat, renamed Irian Jaya by President Suharto in 1973, was later renamed Papua in 2000 under President Abdurrahman Wahid in response to the demands of Papuan nationalists. In 2003, the province of Irian Jaya Barat, later renamed Papua Barat (West Papua), was created. Many Papuan nationalists use the term West Papua or Papua Barat to refer to the entire former Dutch colony of West New Guinea. West Papua, like Indonesia, was a Dutch colony but it did not become part of the newly independent Indonesia in 1949. Against the protests of West Papuans, the UN approved the New York Agreement in 1962, which allowed the territory to move from Dutch to Indonesian control following a referendum. However, by 1963 Indonesia had already assumed control over West Papua. Indonesian President Suharto warned the West Papuans that voting against integration would be an act of "treason." Instead of a direct ballot 1025 local officials were "selected" to vote from a population of 816,000. Prominent West Papuans likely to protest were placed under detention. The USA considered it "necessary" to maintain support for Suharto's Indonesia during the Cold War to stop the spread of Communism.

Alarmed by the activities of the separatist OPM movement and mindful of the fact that Timor had been lost to Indonesia because the Timorese wanted independence the government of Indonesia identified the Grasberg mine as a "vital national asset" and assigned military and police to the site. In 2009, a series of shooting incidents targeting company personnel, contractors, and host government security personnel occurred within the PTFI project area primarily along the remote access road and east levee; these shooting incidents have continued on a sporadic basis. From the beginning of 2009 through mid-February 2012, there were a disturbing 18 fatalities and more than 60 injuries from shooting incidents within the project area. Due to the heightened security situation at PTFI, during 2011 around 1200 host country security personnel (police and military) were assigned to the PTFI project. In addition to the public security presence, PTFI employs around 760 unarmed security personnel and 220 unarmed private security contractors and transportation/logistics consultants. This increased security presence has been necessary to enhance protection of company employees, contractors, and assets and had to be accepted by Freeport as an obligation under its COW agreement with the government.

Inevitably, PTFI was associated with the sometimes harsh measures adopted by the security forces against suspected OPM sympathizers. The complexities of the COW agreement with the government of Indonesia were not well understood by the international critics of mining, and Freeport received a great deal of uninformed criticism about its supposed human rights record. These accusations began in 1994 and continued for the next ten years. Since 1995 a number of independent human rights investigations have been carried out. Those undertaking investigation included the Australian Council for Overseas AID (AFCOA), OXFAM Australia, and the World Development Movement. Indonesian investigations were carried out by Catholic bishops and by the human rights commission, Komnas Ham. These five investigations failed to uncover any irregularity.

## Tailings: Rubbish or an Asset?

Should Freeport consider staking its claim to ownership of the tailings in order to protect its property but also to retain the right to use the tailings or extract the value they represented? Although from the air the lowlands tailings deposition area looks like a World War I battle field there has been a surprising amount of development on the tailings, which were no longer seen as waste. Grass has been planted, cattle are being raised and slaughtered, and vegetables and fruit are being produced. There is a butterfly farm. In the rivers below the airport there is a major fishing operation for barramundi and other fish, which supplies local markets and all the Freeport messes.

The tailings produced very good concrete which Papua badly needed for bridges and roads. The Provincial government in Jayapura wanted to get access to the tailings in order to produce industrial quantities of concrete. Freeport began to consider that it might be important to change the classification of tailings as industrial waste because drilling and test pits in the tailings indicated that magnetite, silver, and copper as well as gold were present in quantities that justified a small-scale mining operation. Magnetite could be sold to coal miners who use it to wash their product. Silver and copper were also present with Chinese investors busy sending samples home and the level of activity suggested the need for a policy on tailings management and a mining plan. This would cover mining methods, footprint, technology, facilities location, and so on and when combined with the information would help PTFI to assess what kinds of business organization and sizes of organization should be encouraged. It would then be possible to estimate how many people could gainfully be employed and whether this effort could be managed by the panners themselves or others. Mining the tailings could introduce a scale of operation which might be copied or emulated by the artisanal miners. This might produce significant changes in the way artisanal miners were organized and reduce the labor-intensive nature of artisanal mining.

To address the larger issues raised by panning Freeport might at some point consider establishing a legal authority such as a River Management Board to manage and coordinate all these complex issues related to river use. Experience suggested that improvement in river management was unlikely if the problem was left to a single department at jobsite. Dealing with the panners and the commercial interest in various uses of the river would require ongoing input from a number of operational areas including, but not limited to, security, health, communities, and engineering. It would also require much better coordination with key stakeholders such as local government and the security forces. Were PTFI to decide to establish a River Management Authority Board it could comprise senior PTFI personnel with members from relevant departments with an elected head who would be responsible for ensuring that all programs activities were coordinated in an open, transparent manner. This would ensure that the activities that threatened the integrity of the property—threats to levees and/or pipelines that carry pyrite and so on—would be effectively embargoed. The board would also have the mandate to consult and invite additional outside stakeholders for consultation, input, and assistance where deemed necessary. In this manner coordinated short-term/long-term goals could be planned and budgets and resources allocated.

## REFLECTION

The Paracatu and Freeport experiences with panning suggest that harnessing social forces may be more rewarding than trying to turn artisanal miners into an industrial organization based on the economics of the firm. Many improvement suggestions assume that artisanal miners can be persuaded to become more efficient and effective, more safety conscious, and more responsible about the impact of their activities on the environment. Yet artisanal miners may well be more responsive to the demands of social, rather than industrial, organization. Social organization is expressed and substantiated by the social relationships the artisanal miners form with each other and the values and beliefs embedded in those relationships. Social organization relies not only on shared values and beliefs but also on the leadership that is created to impose discipline and to provide direction for their work. Social organization provides protection and social security for individuals and families. Panners often enjoy a monopoly in which they are price takers rather than price givers. Unlike workers in an industrial organization they are not greatly concerned with competition or with the need to increase efficiencies in order to survive and prosper. They have no pressing need to upgrade their technology, to alter their division of labor, or to obtain new resources to exploit.[10]

## CLOSURE

Although mining companies have considerable experience with starting mines they have much less experience with closing them. Closure plans need to gain acceptance from a number of audiences, including the local community. Closure presents mining companies with dilemmas about the future of their mine site and its facilities as well as their employees and all those who have come to rely on the wealth created by the operation. Miners cannot expect to resolve these dilemmas on their own. Closure needs a company that has been wise enough to develop the capacity to engage with communities, regulators, government development agencies, politicians, NGOs, and potential investors many years before closure is due to take place.

A number of mining companies have closure regulations and standards which assume the major closure task to be one of making sure the site was safe while ensuring appropriate remediation. But reputation, legacy, local approval, and a sense of neighborliness and decency require much more than that. What is to become of the mine workers and the communities that have relied on the wealth generated by the operation? This task requires imagination and creativity beyond regulation and standards. If employment and well-being have to be secured by means other than mining, then the closure process may need commercial imagination and a light-bulb moment. Communities can be expected to set the bar for leaving high and if they are not happy about the way a company proposes to close then it is likely that the regulators may follow suit.

Integration with ongoing community functions is required rather than the assumption that closure is a special task requiring a plan and an occasional outing like a fire or safety drill. Since two-way communication will be essential as well as the retention of trust and mutual respect it makes sense to assume that community relations has a major role to play. Closure requires intensification, rather than a slackening of community relations. The updating of the baseline will need new information; consultation must be quick and inclusive.

Mining provides technical skills that can be used by other industries and organizations to power the rest of the economy. In PNG, for example, if you can find a plumber or an electrician in Port Moresby it is highly likely that they will have been trained at the Bougainville mine. Companies need to keep a record of the training that has been provided and skills formation that has been achieved as well as the scholarships provided for study both in-country and abroad.

## FLAMBEAU

The fact that the social and community issues that can create major delay and increased complexity in opening and closing a mine was underlined and overcome by the Flambeau Mining Company, a subsidiary of Kennecott Copper, which started and closed a gold mine in Wisconsin within four years between 1993 and 1997.[11] The direct community impacts were minimal though there were concerns about the impact of the mine on water resources as well as the impact that mining might have on wildlife habitats. The ore was rich enough to ship directly to the smelters without having to be concentrated, thereby avoiding the need for construction of a tailings disposal facility. In addition, the company proposed to completely fill the pit with waste rock stored on the surface during the mining operation. After a three-year process involving baseline data gathering, negotiations, and project design and review, the Department of Natural Resources issued the Final Environmental Impact Statement for the project in 1990. A permit hearing was held amid a high level of public input and controversy. However, all permits were issued and construction at the project began in 1991. Ore shipments from the site began in 1993 and continued for more than four years. Backfilling of the pit took 18 months and reclamation activities at the site were completed by the end of 1999.

## RIDGEWAY, SOUTH CAROLINA

Rio Tinto's Ridgeway Mining Company completed gold-mining operations in November 1999. The mine, located in Fairfield County, approximately 20 miles north of Columbia, the capital city of South Carolina, was an open-pit, precious metal mine which produced a doré bar composed of approximately 60 percent gold and 40 percent silver. The Ridgeway Mine operated for 11 years, following early skepticism and outright opposition from local communities. Wealthy neighbors of the mine feared that mining would pose a threat to their way of life as a result of environmental damage caused by the use of cyanide and groundwater pollution. Moreover, closure was a sensitive issue for mining companies in South Carolina because a number of gold-mining companies had declared bankruptcy in order to avoid the costs of a cleanup.

Ridgeway had an outstandingly competent Managing Director with a feel for community relations and from the start of the closure process he wanted the local community and others who would be affected to participate and to share their ideas as to what would work for the mine yet also be acceptable to the community. Mutual support and trust were gradually established by the active engagement of mine opponents and local community groups in the form of information sharing and regular site meetings to discuss plans and issues of concern. Delivering on community commitments was an essential element in the development of trust. This build-up of trust within the local communities, along with their input, helped the company to develop options for future sustainable site uses. Over the operating life of the mine Ridgeway was an active supporter of local education. It provided performance-based awards to local teachers and contributed to state-wide scholarship funds. The construction of a 4-mile water main pipeline linking the mine site to the town allowed local citizens the opportunity to connect to communal water supplies, an example of more substantive community support. There were plans for the town of Ridgeway to extend the water service pipeline to additional customers. Ridgeway also made contributions to the purchase of a community ambulance and other emergency service items.

The Ridgeway Mine closure plan had a regulatory requirement to ensure the long-term physical, chemical, and ecological stability of the site. Achieving and exceeding this objective was seen as fundamental to developing future alternative site uses that would be of benefit to the local communities. With this in mind, Ridgeway carefully reviewed all aspects of the reclamation plan and set out to enhance specific components in order to provide greater levels of site stability. Of prime importance was the stability of the reclaimed tailings impoundment surface cover and creation of wetlands for surface water runoff management and control.

During reclamation presentations to the local communities, one option for future sustainable site use materialized. A local group of educators saw opportunities for the site to be developed as a facility to provide extra-curricular outdoor activities for local school children. The focus would be on fitness and health, environmental education, and traditional cultural value systems. In October 2002, Ridgeway signed a memorandum of understanding (MOU) with the Southeastern Natural Sciences Academy (SNSA) and in November 2003 entered into a license agreement to create a sustainable development environmental research and education center—named the Center for Ecological Restoration. The objectives of the center are to promote a sustainable program for economic growth at the site, balanced with environmental protection and education, achieved by a transfer of knowledge through workshops and seminars coupled with general public interaction. The agreement created a Community Advisory Committee intended to assist and advise an Operations Committee in guiding future decisions and programs of the center. The early focus would be partnering with local universities to establish certified curricula on the graduate/postgraduate level and offer a variety of specialized environmental science courses having applied research requirements. The long-term educational opportunities would involve the local school districts and offer K-12 courses and science field trips covering ecology and the environment. The partnership between SNSA and Ridgeway means sustainable use and educational opportunities for the future of the reclaimed mine site. The future potential of the sustainable development environmental education and research center at Ridgeway seemed to be without limits.

The Managing Director reorganized the mine in the light of these developments with the result that mine employees began performing their post-closure roles for more than a year before closure. When all processing facilities had been removed, the Ridgeway administration and maintenance buildings would remain to provide opportunities for future sustainable development at the site. The buildings total 22,000 square feet of mixed industrial and office space. The regulatory requirement of 30 years of post-closure environmental site monitoring, on and around the 900 acres of previously disturbed land, influenced assessment of the early use of the facility assets by the community. Opportunities for future use of the buildings were further evaluated, engaging various state and local business development organizations, and it quickly became clear that the location of the site was at a considerable disadvantage for industrial development owing to the presence of significant quantities of available, unleased business space immediately adjacent to the Interstate 77 highway corridor.

## ANGLESEY ALUMINIUM METAL

In 2000 it began to be clear that a Rio Tinto subsidiary Anglesey Aluminium Metal (AAM) would have to close within the decade because the cheap electricity supply contract that made smelting possible would not be renewed by the Wylfa nuclear power plant. Without a cheap source of power, aluminum smelters, even those using the latest technology, cannot be profitable. AAM consumed almost 10 percent of the electricity generated in Wales and a smelter at Tiwai Point in New Zealand was estimated to consume almost 50 percent of that country's power production. Temporary subsidies, which had been discussed over the course of the next decade, simply could not make smelting a long-term possibility.

The site near Holyhead, Anglesey's biggest town, was originally selected in the late 1960s because it was a large piece of land which was available for redevelopment in an area with high levels of unemployment. The site also had a deep-water port that provided docking facilities for raw materials and finished product. In 1969 AAM bought Penrhos Park and opened it to public access. From 1971 to 2009, the AAM smelter was one of the island's biggest employers. A Rio Tinto–managed joint venture with Kaiser Aluminum, the plant employed up to 570 staff and 70 full-time contractors, and had a capacity of 145,000 tons per year of metal.

The environmental measures that would be required by closure were not extensive. In the early days of the smelter local farmers had complained that chemicals emitted from the plant were harming their sheep but this had been overcome by regular consultations with veterinary surgeons. However, rather than getting on with preparing the community for closure and looking for alternative uses of the plant, AAM management kept the looming disaster to themselves by insisting that there might be a

White Knight solution. Instead of intensifying the capacity to engage with the community, use of the C word was discouraged. For over a decade little was done to prepare the community for the reality of what closure would mean and as a result the reputation of the company and the future of families in the community were put at risk.

AAM management did not seem to be able to move beyond smelting or some other form of manufacturing that would result in minimal change. The most community-oriented thing that AAM did was to ensure that those who telephoned the plant were greeted in Welsh. AAM management showed little interest in using community relations as one of the key weapons in their approach to closure. The social baseline was outsourced to a local university and those doing the analysis did not really comprehend the closure issues. The five-year plan was no better, and it dodged the closure issue. Closure required, but did not get, enough managers with an aptitude for community relations and people who have the experience and the skill set to be able to understand and pursue solutions with communities, government agencies, and politicians, all of which require imagination and creativity. Making the community part of the process of developing alternative possibilities would have increased the robustness and acceptability of the process and outcome while avoiding any sense that AAM's closure process had lacked community transparency.

When it became known within Rio Tinto that AAM would probably close, the very best people began to leave the refinery as soon as they could. Being posted to Anglesey was not seen as a stepping stone to greatness. One excellent manager said somewhat bitterly as he was leaving for a better job elsewhere that "Anglesey was where pebble dash goes to die."[12] With prodding from London, studies were done to help identify the impact of closure as well as measures that might mitigate that impact. For example, examination of the age profile of employees showed how many were close to retirement. However, several had mortgages to pay off and children to educate. Many AAM employees who had children did not want their sons and daughters to have to move away from Anglesey in order to make their living.

Despite the intense efforts of AAM and union and government representatives, the smelter was unable to secure another commercially viable power contract. In September 2009, smelting operations ceased, with the loss of 400 jobs. Part of the site became a remelt facility, preserving jobs for almost 100 people, but this too was decommissioned in 2013 owing to the cost of raw materials. However, it had been clear as early as 2000 that the nuclear power station would not extend the power supply contract.

The redundancies of 2009 sparked a bitter and angry community backlash against AAM perhaps in part because the community had not been brought along and included in the company's thinking. Mine managers had kept their cards close to their chest because they felt it would be dangerous to appear to take the idea of closure seriously. For almost 10 years AAM held to this belief. Instead of being energetic and being seen to go out to develop options in consultation with the local community, AAM seemed to be waiting for potential suitors to approach them.

AAM had three assets: the port, the Penrhos Park and its associated land, and the plant where smelting took place. Shippers and tourist agencies went to AAM with proposals for using the port facilities. Eventually a proposal to use the smelter site with the potential to create up to 1000 jobs emerged. At its heart will be a 299MWe biomass power station. Output from this plant will drive the rest of the development, which will include hydroponic and aquaculture facilities producing vegetables and over 8000 tons of fish a year, plus a plant producing compostable food packaging. A processing center will produce value-added products from food produced and packaged on site. Planning permission is already in place for the 299MWe Biomass Energy Center, and Lateral say they expect to be fully operational by 2017 with 500 permanent staff. Friends of the Earth have been campaigning against this proposal on grounds that it will not produce green energy. Good closure is not a matter of jobs lost and jobs gained, but of trying to make sure that any new jobs are suited to the skills and experience of the mine workers. The second project is to be developed by Land & Lakes, a company which is planning a tourism, leisure, and housing venture on Penrhos Park land. This development aims to develop holiday and residential accommodation.

## KELIAN

In July 1985 the Kelian Equatorial Mining Company (KEM), which was owned by Rio Tinto, signed a COW agreement with the Indonesian government and started mining on 1 July 1991 producing between 13 and 15 tons of gold a year. Mining ceased in 2004 and, after working through the stockpiled ore, the company intended to rework some of the lower-grade stockpiled ore before leaving the site in 2007. KEM management assumed that the main task for closure was to secure regulatory approval for the steps to be taken by the closure process from government, the community, and civil society. To develop a closure plan a Mine Closure

Steering Committee (MCSC) was established with representatives of KEM, Rio Tinto London, local government and central government, the local Customary Council, and an organization called LKMTL which had been set up to look after special community concerns.

Four working groups were asked to study the tailings deposition areas and its dams, environmental remediation, the disposal of physical assets, community empowerment, and regional planning. The MCSC met on a quarterly basis for several years. Members discussed suggestions raised by the community and the authority. The Forestry and Lands Departments from the government were concerned because of the need to care for the tree cover for the watershed that would protect the impounded tailings. The Department of Mines, which had statutory responsibilities for closure, was interested in seeing how the closure plan was developed, since Kelian was the first large mine to be closed and they wanted to use the Kelian experience as a basis for making closure regulations. Local government in the shape of the Bupati and nearby communities were interested not only in the safety of the site but also what was to become of the site and its assets in the future. The World Bank sent representatives who helped the MCSC to understand the economic effects of closure on the local economy. Community and economic development issues were able to draw on the experience of the Kelian Foundation. The Foundation, which was well staffed and funded, had an excellent track record in delivering education and health services, including clean water for villages and expert agricultural advice from Rio Tinto's only professionally qualified tropical agriculturalist.

The MCSC was able to get agreement on the environmental aspects relatively quickly. The impounded tailings and waste dumps had to be rendered safe and the area either re-vegetated or turned into managed wetlands. Waste water from the Nakan dam would be drained into the mine pit and, from there, into a 30–40 hectare "wetland system" that would eventually reach the Mahakam river. KEM were confident that this system would neutralize any toxic chemicals. Each waste dump would be confined within a retaining dam. Waste rock and other heaps which did not contain toxic materials would also be dealt with by the "dry cover" method. These heaps would be covered with clay and topsoil and then planted. The watershed area containing the tailings had to remain forested so that the tailings impoundment area was protected. To do this KEM proposed to employ rangers to prevent tree-cutting and squatting. Security was also needed to preserve the integrity of the containing structures for the tailings impoundment area.

Ten years after mining ceased Kelian has still not been able to formally close. In many instances when mining ceases closure may not necessarily turn out as expected when success is thought to be a matter of satisfying environmental regulations. While the environmental proposals reflected state-of-the-art thinking it was less clear that they reflected what local people wanted. Beginning in the late 1990s the local river on the property had been invaded by hundreds of panners who were working 24/7 to capture gold in the tailings with the aid of air hoses and lighting powered by generator sets. What KEM proposed to do to remove the artisanal threat was mine the area so that the artisanal miners would see that there was no gold left. However, there was a rumor that one of the best spots for panning was under KEM's mess hall. Was it really realistic to assume that local people who knew that there was still gold in the waste dumps and impoundment areas would leave the area undisturbed?

The MCSC did not do so well when it came to deciding what to do about the facilities, buildings, plant, and so on that would no longer be needed when mining ceased. The local Bupati who had been a transport superintendent at the mine tried to have the physical assets gifted to his administration. Then he suggested that the site could be turned into a casino. More promising was a suggestion from a foreign university to turn the mine's facilities into a wildlife educational facility.

The MCSC did make a serious attempt to provide for local employees after closure although surveys showed that many wanted to retire on their retrenchment money. Here the World Bank was helpful in making suggestions about how to avoid inflation in the price of land and property. It was important to make sure that the recipients of large sums did not waste their money on gambling and loose living. Slightly worrying was the fact that a number of those to be retrenched wanted to be rubber planters. However, producing rubber is extremely hard work. With respect to agriculture, Kelian's tropical agriculturalist was able to assess the soils in the region and provide advice and training to would-be farmers. His work was so successful that when mining companies visited the area they were unable to find any recruits.

KEM had to set aside funds to pay for closure and an offshore fund was located in Singapore. However, just as it began to look as if the closure process was done and dusted new problems emerged. Closure represents the last chance for those who believe they have a claim against the company. When KEM had acquired the site from the government of Indonesia the company assumed that it could go ahead and mine because the central government would have taken care of any local liabilities. Not so. What KEM had not sufficiently appreciated was the fact that a number of artisanal miners had been panning in the river before mining ended. These artisanal miners began to make representations to KEM that they were entitled to compensation at the level of up to 10 grams of gold a day for several years. That group of claimants was joined by another group of angry local people who claimed that they were entitled to compensation because their homes had been destroyed in order to make way for the KEM operation. Each of these claims had to be examined in some detail to establish which of several types of house the claimants had been occupying.

The claimants exerted pressure on Rio Tinto, preventing the finalization of closure. In June 1998, KEM had signed an agreement to negotiate with the community organization LKMTL, following community demands presented at annual shareholders' meetings in London and Melbourne. While the gold panning and artisanal claims could, with time and effort, be processed KEM was in more serious difficulty over human rights allegations. Early in KEM's operation expatriate employees had engaged in sexual relations with girls as young as 12 years of age. However, instead of the perpetrators being turned over to the authorities they were given their passports, which were kept by mine management during their stay in Indonesia, and allowed to return to their home country. When Rio Tinto London realized the scale and seriousness of the wrongdoing an Australian federal court judge, Marcus Einfeld, was asked to investigate, negotiate, and reach a settlement which was acceptable to all. This was done but later there was a bizarre postscript to Einfeld's involvement.[13]

## Notes

1. Tailings consist of ground rock and process effluents that are generated in a mine processing plant. Mechanical and chemical processes are used to extract the gold or other metal product from crushing the ore and produce a waste stream known as tailings. This process of extraction is never 100 percent efficient.
2. See Penna-Friema and Bredgo (2007).
3. See Hanai (2000).
4. See Heemskerk (2005), Davidson (1993).
5. See Scott (1997, 2011).
6. See Leith (2003), Mealey (1996).
7. This account is based on a 15-year association with PT Freeport Indonesia (PTFI). During this period the leaders of the function were Charlie White,

David Lowry, and then Stan Batey. Stan, an exceptionally gifted professional, died suddenly and tragically in 2016.

8. Over 400,000 Dani lived in the Baliem valley in the highlands. See Heider (1970). Robert Gardner made a film called "Dead Birds" about the Dani as part of the 1961 Peabody Museum expedition. The film's title is borrowed from a Dani fable that there was once a great race between a bird and a snake, which was to determine the lives of human beings. Should men shed their skins and live forever like snakes, or die like birds? The bird won the race, dictating that man must die. The Asmat, a fierce coastal people, lived on the coast less than 100 kilometers from PTFI. They were well-known cannibals who were suspected of killing, and perhaps consuming, Michael Rockefeller, son of New York State Governor Nelson Rockefeller, when he was a member of the 1961 Peabody Museum expedition to the area.

9. See Labonne (1996).

10. See Conner (1991).

11. A coalition of First Nations, the Sierra Club, anglers, and wildlife enthusiasts prevented the development of another small mine in Wisconsin. See Lynch (2014).

12. *Pebbledash* is a coarse plaster surface used on outside walls that consists of lime and sometimes cement mixed with sand, small gravel, and often *pebbles* or shells. The materials are mixed into a slurry and are then thrown at the working surface with a trowel or scoop.

13. Former Australian Federal Court judge, civil libertarian, human rights advocate, and Living National Treasure Marcus Einfeld OA (since rescinded) made a big mistake when he was caught speeding in Sydney. He claimed his car was driven at the time by a US-based Australian friend, feminist philosopher, and author Teresa Brennan. In fact Brennan had died three years earlier, the victim of a mysterious and suspicious hit-and-run accident. Instead of confessing, Einfeld compounded his lie by then claiming that another Teresa Brennan, also US-based, also deceased, was driving his car. The resulting, highly publicized, court case ended with Einfeld being convicted of perjury and intending to pervert the course of justice. He was sentenced to three years in prison but was released on parole after serving two, his reputation in tatters.

## Bibliography

Conner, Kathleen R. 1991. A Historical Comparison of Resource-Based Theory and Five Schools of Thought within Industrial Organization Economics: Do We Have a New Theory of the Firm. *Journal of Management* 17(1): 121–154.

Davidson, Jeffrey. 1993. The Transformation and Successful Development of Small-Scale Mining Enterprises in Developing Countries. *Natural Resources Forum* 17(4): 315–326.

Hanai, Maria. 2000. Formal and Garimpo Gold Mining and the Environment in Brazil. In *Mining and the Environment: IDRC Case Studies from the Americas*, ed. A. Warhurst. Ottawa: Government of Canada.

Heemskerk, Marieke. 2005. Collecting Data in Artisanal and Small-Scale Mining Communities: Measuring Progress Towards More Sustainable Livelihoods. *Natural Resources Forum* 29: 82–87.

Heider, Karl. 1970. *The Dugum Dani: A Papuan Culture in the Highlands of West New Guinea*. Chicago: Aldine Publishing.

Labonne, Béatrice. 1996. Artisanal Mining: An Economic Stepping Stone for Women. *Natural Resources Forum* 20(2): 117–122.

Leith, Denise. 2003. *The Politics of Power: Freeport in Suharto's Indonesia*. Honolulu: University of Hawaii Press.

Lynch, Larry. 2014. History of the Crandon Mine Project and Implications of Lessons Learned. SME Wisconsin Annual Conference on Partnering for Sustainable Mining: Hard Rock and Soft Rock Mining in Minnesota and Wisconsin, Eau Claire, Wisconsin.

Mealey, George. 1996. *Grasberg—Mining the Richest and Most Remote Deposit of Copper and Gold in the World in the Mountains of Irian Jaya, Indonesia*. Singapore: Freeport-McMoRan Copper & Gold Inc.

Penna-Friema, Rodrigo, and Eduardo Bredgo. 2007. The Risks of Commoditising Poverty: Rural Communities, Quilombola Identity and Natural Conservation in Brazil. *Habitas, Goiania*, vol. V, July.

Scott, Parry. 1997. Garimpos, Community and Rio Paracatu. Unpublished Report.

———. 2011. Families, Nations and Generations in Women's International Migration. *Vibrant, Virtual Brazilian, Anthropology* 8(2): 279–306.

# PART III

# Corporate Social Responsibility

# CHAPTER 8

# Data and Forms of CSR

Aid agencies and NGOs have used Corporate Social Responsibility (CSR) to assess the relationships between mining companies and communities but without first collecting baseline data or engaging with communities or building hands-on skills. In these circumstances, CSR has made it seductively easy for these organizations to make judgments on community situations from the comfort of their offices. However, social responsibilities are created by society, not by aid agencies, NGOs, or mining companies. In most societies social responsibilities are predominantly voluntary and established by custom and convention.[1] CSR assigns responsibilities to mining companies without any attention being paid to the opinion or the rights of the company or to the norms and values of society, voluntarism, or reciprocity. To substantiate a claim to being a social process CSR should be able to show evidence that there are values and beliefs related to that responsibility that are shared by the community and the company: community responsibilities toward the company would also have to be covered. CSR would have to show that there are well-established relationships between the company and the community that generate, validate, and sustain these concepts of social responsibility and all of this would obviously have to be substantiated by a good social baseline.

Unfortunately, and despite its popularity, CSR has had too little to do with what society thinks of as being social. Take for example, The World Business Council for Sustainable Development's (WBCSD) statement that

© The Author(s) 2017
G. Cochrane, *Anthropology in the Mining Industry,*
DOI 10.1007/978-3-319-50310-3_8

"Corporate social responsibility is the commitment of business to contribute to sustainable economic development, working with employees, their families, the local community, and society at large to improve their quality of life."[2] This is not very social, nor is the WBCSD's listing of key issues identified in its first 1999 CSR report: human rights, employee rights, environmental protection, community involvement, and supplier relations.

Companies have social responsibilities that should be maintained in good times, and in not-so-good times, throughout the life of a mine. They have a social responsibility to ensure that their employees have, or acquire, the language and the hands-on skills to be able to earn and maintain local understanding and respect. Companies have a responsibility to pass on to the community information which may affect its way of life and related to the operation, for example, expansions or threats to profitability. They should report on how mining is affecting the physical environment, air, and water resources. They have a responsibility to maintain open and transparent channels of communication and to ensure that their senior management is on top of the detail of local community relations and is keeping up to date with local affairs issues and events. To do these things companies have a responsibility to maintain social relationships in the community, and they should establish and maintain a cadre of company helpers with the capacity and the willingness to live close by or in the communities they want to assist.

Communities have social responsibilities including those for the definition and preservation of their cultural heritage, for preventing damage to their environment, safeguarding food security, and when they are beneficiaries under a mining agreement to use the wealth generated by the mine that they receive for the benefit of all and for future generations; they also have responsibilities for representing the relationship that they have with a mining company in a fair and objective manner to third parties; and they have a responsibility to use their community institutions to maintain an open and transparent dialog with the company.

CSR is a social fiction in much the same way that the idea of the corporation as an individual is a legal fiction. But CSR is not well constructed. CSR does not make social sense in the way that the legal fiction of the corporation makes sense. Commerce is facilitated when corporations are able to hold property, and to sue and be sued.[3] The corporation has legal rights which it can pursue in court as well as responsibilities. This is not the case with CSRs because reciprocity, which is an essential component of social relationships, is not part of CSR thinking so companies have obligations and responsibilities but no social rights.

While it is clear that mining companies no matter where in the world should be held to account for their behavior in the communities into which their operations intrude it must also be reasonable to suppose that all those who intrude into the social life of communities, including NGOs and aid agencies, should accept that they have social responsibilities.[4] But what sort of responsibilities? How are these responsibilities determined? For companies and communities CSR will work best when seen as a social process where both sides get to know each other.

Common sense suggests that for CSR to be useful those defining the concept must have local skills and knowledge if they are to make a valid interpretation of a company's CSR. Sensible people should, without difficulty, be able to agree that deciding what a mining company's CSR is in a remote traditional community thousands of miles away does not make sense. Paracatu Mineração in Brazil asked local communities to tell them what social responsibilities the company should pursue and provided a list of items that they thought would be useful to the community. The local view was that these activities would be welcome but they were not seen as the social responsibilities of the company. A number of the proposed initiatives were declined on the grounds that the company did not have the right skills and experience to do the work and in other cases they said they had identified a donor who they thought could do the task better than the miners. They asked instead if it was possible for the company to help with the cleanup of damage caused to nearby streams and rivers by artisanal mining. Not all communities are so reasonable. In Indonesia a mining company queried a demand that local wages be brought up to US levels, the same level as a company was paying in the USA, and was told to consider the increase part of their CSR!

CSR assessments should be supported by an appropriate data base but the information UN agencies and NGOs have compiled does not usually capture the relationships between people and mining companies in communities. An assumption that the benefits of economic growth might trickle down to communities might not be accurate. Incomes might be rising while local society was becoming fragmented. Constructive exchange between those who like to work in communities and those who believe in global approaches is not common because those interested in the worm's eye view want to burrow more deeply into the social soil and those who believe in the bird's eye view want to rise higher to get a clear view and to be better seen.

## CSR Based on Fieldwork

Rio Tinto's social baseline used fieldwork methods which relied on the anthropologist in the community collecting the widest possible range of data in order to have generalization derived from the facts that have been collected. The anthropologist has to spend months or years on the ground getting to know the people, their language, their customs and traditions, the way they organize, and their beliefs, values, and attitudes. When fieldwork is used to understand social responsibilities the facts on the ground suggest the appropriateness of the conclusions and contrary facts have a chance to appear and to be considered.[5]

### *Free Prior and Informed Consent*

The FPIC responsibility concept has been developed on the ground in cooperation with indigenous people. Prima facie free, prior, informed, consent should be treated with respect because indigenous peoples have adopted it as their protocol to be used in their dealings with miners and other developers. The importance of FPIC compliance is now recognized by the Organisation for Economic Co-operation and Development (OECD), International Labor Organization (ILO), Food and Agriculture Organization (FAO), United Nations Environmental Program (UNEP), and non-governmental organizations (NGOs).[6]

FPIC acknowledges the importance of a mining company and a community getting to know each other and working hard to see if they can develop a mutually satisfactory relationship. A number of governments and mining companies have attempted to use FPIC to force a here-and-now decision on the part of the community before they have had time to get to know the developer. In these circumstances FPIC operates like a traffic light put up by well-meaning people who want to prevent traffic accidents.[7] All approaching traffic is shown a red light. Yet surely this approach should be seen as being unhelpful because it does not facilitate a process of listening and learning before engagement[8]; instead it starts by discouraging engagement. This sort of use of FPIC does not suggest to a mining company and a remote community that it might be useful to have a relationship before deciding to move forward or end contact. Given the ups and downs of the project cycle forcing a here-and-now decision does not make good social or business sense.

Both the Ridgeway and Diavik Diamond mines followed the FPIC process; they initiated and maintained a process of engagement which in the case of Diavik will carry on through the life of the mine. Using FPIC in this way FPIC is entirely consistent with the approach to community relations established by Rio Tinto. The FPIC engagement process needs solid baseline information and intermediaries who know the community and who are trusted by that community to explain and give advice. In short, FPIC provides a blueprint for good community relations.

The traffic light use of FPIC may deny some communities the opportunity to develop a project that could provide useful benefits. For example, many indigenous groups have entered into agreements with energy companies and the state over the use of their land for extraction purposes. O'Faircheallaigh[9] says the incentives for such a decision are the prospect of better economic opportunities for younger generations, which would also counteract problems such as substance abuse, imprisonment, domestic violence, and suicide that derive from the loss of a sense of belonging. He calls it a "balancing act" between people's desire for development, their desire to look after country, to benefit financially, and to be included in an otherwise rather antagonizing society—the "settler" society.

Oxfam's Australia FPIC Guide says FPIC requires:

> **Free** from force, intimidation, manipulation, coercion or pressure by any government or company; **Prior** to government allocating land for particular land uses and prior to approval of specific projects. You must be given enough time to consider all the information and make a decision; **Informed**, you must be given all the relevant information to make your decision about whether to agree to the project or not. Also: This information must be in a language that you can easily understand. You must have access to independent information, not just information from the project developers or your government. You must also have access to experts on law and technical issues, if requested, to help make your decision; **Consent** requires that the people involved in the project allow indigenous communities to say "Yes" or "No" to the project and at each stage of the project, according to the decision-making process of your choice.

The right to give or withhold consent is the most important difference between the rights of indigenous peoples and other project-affected peoples.[10] Progressive community relations thinking should see that FPIC, and its identification of the way outsiders have to work with indigenous peoples in order to develop mutual understanding and trust, can represent an advance for community relations. Can a mining company like Rio Tinto propose to follow a higher standard of preparation, analysis, and care for indigenous peoples than for other communities?

## Social Contracts

A good deal of what mining companies do with their community relations can be seen as a derivative of social contract thinking. The idea of a social contract has an ancient lineage going back to Rousseau and contract thinking is at the heart of FPIC.[11] The social contract depends on men and women making a calculation that, in the main, they will be better off with the contract than without it. Social contract thinking is a bargain that must be underpinned by trust. Hobbes and Locke were realistic; realizing that to have a bargain and a government that can last long enough to work it was necessary to have that trust because the future is unknown and, while one can agree to a government now because what it is doing is popular, this may not continue to be the case.

Like a marriage contract, which is for better or worse, a social contract is not expected to forever please all who have agreed to it. For Hobbes and Locke there could not be a bargain, certainly no bargain that could last unless there was some recognition and acceptance of the need to take the rough with the smooth. Social Contract bargains have to be made for the common good and will not sit well with everyone. It is a cardinal requirement of collective living that some individuals will always have to make sacrifices for the common good. If the government bans smoking some will be unhappy.[12]

Promoting items that only appeal to a small section of the population will make it difficult to assume the existence of a social contract. If narrow fine-grained agreements about social license have to be seen as part of the bargain, it would become much more difficult to assume the existence of a contract because common sense tells us that when a narrow bargain is introduced some are likely to be in favor and some not.[13]

## CSR BASED ON "IF I WERE A HORSE" THINKING

Aid agencies and NGOs who have not undertaken fieldwork have to base their assessment of CSR on hunches and assumptions. Those relying on assumptions try to put themselves in the shoes of people they do not know.

This sort of imagining commits what E.E. Evans-Pritchard called the "if I were a horse" fallacy: "Because we are not horses, have never been horses, and cannot know what it is like to be a horse, our speculations about horse-sense probably have little or no connection to horse-reality."[14] A good example of reliance on "if I were a horse" thinking is provided by Rapid Rural Appraisal (RRA). RRA a way of quickly collecting data, used by aid agencies, consulting firms, and a few mining companies, to gain community understanding without having to engage in lengthy anthropological fieldwork:

> Such appraisal involves avoiding the traps of quick and dirty or long and dirty methods and using instead methods that are more cost-effective. To do this means ignoring inappropriate professional standards and instead applying a new rigor based on the two principles of optimal ignorance—knowing what it is not worth knowing—and proportionate accuracy—recognizing the degree of accuracy required ... Time (and cost) was emphasized as a critical factor in effective appraisal and rapid rural appraisal methods increase the chance of reducing the bias against the poorer rural people in the promotion of rural development.[15]

The RRA surveyors who usually have had no prior experience with the community being studied come from a number of disciplinary backgrounds and may have no common training prior to going to the community. RRA assumes that what needs to be known about various communities overseas can be done by questionnaire or even a short survey of a few days, weeks, or even a couple of months. The extent to which this method can collect highly sensitive information is doubtful. The RRA process could not sustain community relations though it could be useful in special circumstances such as when one company wants to buy another and a quick decision is needed.

### Social Impact Assessment

Social Impact Assessment (SIA) can easily turn out to be "if I were a horse" thinking. Does a local concept of impact exist and if so what is it? Is this concept of impact the same as ours? This question is seldom posed and answered, but it really should be because we cannot simply assume that a community has such a concept and that it can be understood in the same way that we understand impact.[16]

The prediction of social change, which is what impact assessment is all about, becomes highly problematic when left to someone who does not know the members of the community and does not have time to get to know them. Social scientists who have not collected social baseline data have no calculus which enables them to answer social impact questions with any degree of predictive authority. It is an unsatisfactory position for a community to be in to find itself in a situation where those in charge of their impact assessment think they know what will be worst or best for a community.

"If I were a horse" thinking is often prompted by assumptions that are not discussed or made public. For example, traditional people have been seen as being in danger, as weak and defenseless and incapable of absorbing, or accommodating to, change. Industrial projects have been seen as (inevitable) destroyers of a unique, and irreplaceable, way of life because the mining industry had a substantial trust deficit. Industry critics began from the position that companies would try to cut corners. Whereas in the rich countries the questions and the answers were supposed to emerge from the objective assessments of community opinion, the investigative agenda in poor countries is often heavily influenced by critics in the industrialized countries.

As a result of SIA's preoccupation with harm too little attention has been paid to assessing the benefits that could come from creating jobs, income-earning business opportunities, or service provision. Too much emphasis has been placed on preservation and conservation of the traditional way of life at the expense of appreciating change requirements that reflect what local people want. In developing countries it is usually impossible to stand still or suggest that no change performs any service for traditional peoples. Rapid population increases—birth rates of 4 percent and more—erode the resource base; increasing knowledge affects preferences, particularly the preferences of the younger generation.

Will change really cause mental indigestion? It would be a mistake to assume that communities with limited exposure to technology will always be overawed by industrial might. When I lived on Guadalcanal in the South Pacific a former US Army stretcher bearer from Elizabethville, New Jersey, told me that each day in 1943 more vehicles traveled down Mendana Avenue in the capital, Honiara, than through any city in Europe. He said local people took the traffic in their stride. The mental indigestion attitude was captured by the anthropologist F.E. Williams in 1939 in an address to the Australian and New Zealand Academy of Science when he said that some considered culture to be like a watch: change one cog and it would not work.[17]

Investors, regulators, and international aid agencies are required to undertake, or see the results of, Social Impact Assessment before they get permission to develop new mines in the developed and developing countries. Social Impact Assessments are also conducted when there is an application to extend operations or the life of the mine. Consultants undertaking social impact assessments frequently went to the field without taking the state of a mining company's community relations into account or sighting the mining company's baseline, consultation guidance, and data about community aspirations.

Despite claims being made to the contrary, SIA is overwhelmingly structured in terms of harm. Those using the concept talk about harm even though they may have little understanding of local ethnography and certainly do not have the time to remedy this deficiency when carrying out an impact assessment over the course of a few weeks. Despite attempts to say that impact assessment is also capable of examining opportunities and positive developments, the universal use of "mitigation" and "compensation" language in the assessment continues to confirm the negativity of this environmentally influenced approach.[18] It is car crash analysis; you can hear the car hitting the body. Consequently it is hard for SIA to generate community-level information of the same quality as is produced by a social relations approach underpinned by a social baseline.

Again, as with FPIC, impact analysis is often used as a traffic light instead of viewing it as a continuing process to let developers and communities get to know each other. An FPIC process would make the impact analysis more useful since at present, beyond the use of the term SIA and some general acceptance of the idea that the analysis should deal with the risks and opportunities presented by a project, there is no agreement on methods or content. An SIA might be five pages or five volumes, long, depending on the personal proclivities of the author. What one analyst sees as a risk may be seen by another as an opportunity. Impact assessment works for environmental issues where there are metrics and black and white issues but it does not work so well when it becomes, as is now so often the case, the principal form of social analysis.

SIA was born in 1970 when the National Environmental Policy Act (NEPA) was signed by President Nixon. Section 102 of the Act required

Federal agencies to make "integrated use of the natural and social sciences in decision-making which may have an impact on man's environment." In the USA, regulators thought that what was needed was the widest possible public discussion and elimination of any suspicion that those with the most education should have the greatest say; SIA needed broad-ranging inquiry, tested in public meetings. In the USA the work often employed a range of social scientists, economists, geographers, psychologists, and sociologists. Impact assessment did not require the "best" or the "right" decision. Nor was there a requirement to stop projects likely to have negative consequences. The emphasis was on public disclosure. Most impact work in the USA insists on a full and frank discussion of likely impacts or effects (along with measures to lessen or mitigate negative impacts that are identified). The work is then usually distributed to interested parties. Impact analysis in the USA was designed to address the "unintended consequences" of developments that are normally initiated by private, profit-oriented, companies. The important point to bear in mind has been that SIAs in the USA have given priority to local opinions. This is because expert analysis has been shown to be no better than local opinion.

SIA is perhaps best viewed, and has its greatest utility, as serving to stimulate and facilitate public consultation and discussion. There is a clear and present danger that if the SIA process does not move toward an emphasis on public consultation the results will continue to award much greater weight to the opinions of the literate, articulate views of the experts facilitating assessment than it does to the opinions, aspirations, and wishes of communities that are being asked to become the neighbors of a mine. Too often the SIA process is a conversation between the developer, the regulator, and critics of mining with local people being little more than bystanders. What is needed is a socially grounded process which allows community people a greater opportunity to generate for themselves the questions and the answers about good and bad change. As time passes, and as SIAs become more embedded in a process of community relations, local people will then be more involved and SIA practitioners will thus have a role more concerned with facilitation than well-meaning guesswork.

Why has there been so little concern for the local economic impact of a mining project among aid agencies and regulators? For many communities this is much more important than the environmental impact. Communities should be seeing an economic, environmental and social impact assessment, but this has not yet been taken on board by the mining associations or the regulators.

## Social License to Operate

Jim Cooney of Placer Dome Inc has been widely recognized as the originator of the term "Social License." Its appearance in the literature and the ascendance of the concept has been important from a scholarly and practical perspective. Susan Joyce and Ian Thompson, who recommended the use of Jim Cooney's idea of a Social License to Operate,[19] said:

> We propose that a Social License to Operate exists when a mineral exploration or mining project is seen as having the approval, the broad acceptance of society to conduct its activities. It is a license which cannot be provided by civil authorities, by political structures, or even by the legal system. Most importantly for the state of current discussions within the mining industry, it cannot be claimed as a product of an internal corporate process such as an audit of company practices. It can only come from the acceptance granted by your neighbors. Such acceptability must be achieved on many levels, but it must begin with, and be firmly grounded in, the social acceptance of the resource development by local communities.[20]

In order to avoid being seen as "if I were a horse" thinking it is essential that the Social License idea be supported by sound baseline data and an established community relations program. It is necessary to see that the facts on the ground support and flesh out what a social license means locally. Is the concept of a license, or something very close to such an idea, locally understood? How was this license applied for and secured? What conditions are imposed and what would be the grounds for cancellation?

Those who promote the idea that mining companies should have to gain a Social License to Operate might do well to consider the requirements for a social license to close a mine. As we have seen with Kelian community approval was seen as being fundamental to a successful closure process. When Rio Tinto did not develop a mine an attempt was usually made during the exit to provide something of social value to communities. For example, when the company decided to exit Panama, where there was no community relations program, a road was constructed to help Guaymi with access to markets and health facilities; at Kelian there was a lengthy closure planning process after mining finished in which communities participated and from which they obtained livelihood benefits; before exiting Borneo a wide range of government officials had been introduced to the requirements for successful resettlement caused by dam construction, and the authorities were left a number of resource materials, including a contemporary ethnography and a methodology for doing further work of this nature. In Guinea the government was provided with a feasibility study for the development of its iron ore resources which contained a great deal of the social material that any future iron ore development would require. However, rather than continuing to apply ad hoc social closure measures it would make sense for Rio Tinto and other mining companies to set down a procedure to be followed in all exits regardless of local closure regulation.

### Legal Agreements

With a Social License and any legal agreement it is essential that the social baseline makes it quite clear who in a local community has the authority to sign such documents. Where this cannot be demonstrated the acquisition may be *ultra vires* from a traditional perspective; it may also provide an example of "if I were a horse" thinking. A second requirement is that *consensus ad idem* exists, meaning that evidence can be supplied to show that both sides are fully aware of what the contract is about. For example, local people may have sometimes believed that they were conveying a limited right to use a piece of ground whereas the company thought that they had bought the land.

Rio Tinto published *Why Agreements Matter* in 2016. This suggested that the process of making a legal agreement with a community on mining or important issues related to mining, would enable Managing Directors to avoid sleepless nights spent worrying about nasty surprises. The inspiration for this approach appeared to come from Australia where legal agreements were necessary as a result of a unique legal history concerning Aboriginal land rights, which meant that mining companies had no choice other than to enter into legal agreements if they wanted to mine on land owned by Aboriginal people. However, other countries, and other communities, had different ways of making land available for mining and a legal approach could be risky in countries where the gap between community and government was large, political stability was wobbly at the best of times, and social change was in the air. In these circumstances there was no substitute for a well-staffed and well-trained community relations team continually on sentry duty. Reaching local understanding can be helpful but not if as is usually the case it uses language, concepts, and content that could only have come from a mining company wish list, or if it requires many lawyers to approve everything. There is a more fundamental objection to flagging agreement-making related to mining as something that communities people may be able to deliver. Earlier in the book it was mentioned that when communities personnel were joined with security personnel their effectiveness was impaired. Imagine the damage that could be done to the function if it were known in the community that it was engaged in relationship building in order to get a mining deal that senior managers wanted. It is always difficult for communities personnel to balance their role as advocates for their community with their role as company men and women; involvement in high-profile agreement-making on the side of the company in situations where few will believe that there is a level playing field will not perhaps be a wise thing to embark on. Surely it would be best to give priority to building the relationships that will enable both sides to get a good deal and trying to ensure that any deal reached can be maintained with community support?

Legal agreements record the way company lawyers view rights, duties, and responsibilities in their own orderly world. This does not require the lawyers to drill down to capture and understand the drivers and the realities of social change or to critically examine how local people see the quality of the relationships that the mining community has with them.

A legal agreement with a government to begin mining will often have little to do with community relations personnel as was the case with the old Contract of Work (COW) system in Indonesia. Agreements to mine may contain details about compensation where appropriate, employment targets and the provision of training, as well as supply chain and environmental commitments. Obviously a mining company will commit considerable treasure and professional time in order to spend a great deal to secure permission to mine. Indeed there are rumors, but figures have not been made available, suggesting that the amount spent to negotiate, draft and finalize iron ore agreements with Aboriginal owners in western Australia could have established several law firms.

Neither the Murowa resettlement in Zimbabwe nor the Diavik Diamond Mine Participation Agreements in Canada relied on legal agreements. Diavik and Murowa depended on the quality of the community relations personnel involved. Rio Tinto's successful community engagement was based on an assumption that understandings would only be possible when sound community relations had been established and there was a team in place which had generated mutual understanding, trust and respect. Setting down just what it is that a community can expect by way of benefits may appear to provide a powerful form of persuasion for wavering or undecided communities. Obviously such agreements have appeal for senior mining company managers, international observers, and critics of the mining industry who are sure that the local people will get a bad deal. However, in the early days of a project too little may be known about the quantum of benefits the mine will generate and, as a result, it will be hard to estimate what the local portion might look like. The benefits produced by a mine may be slow to appear and may be irregular or "lumpy." Several years may have to pass before revenue streams become steady, thus permitting a company to make more accurate assessments about what levels of community support can be provided.

In any mining company the idea of an agreement will attract armies of lawyers and this will distract from the fundamentally important challenge of building close relationships between the miners and their neighbors. What Bougainville demonstrated, and it should not be forgotten, was that it was not making agreements that was important; what was critical was making sure that miners behaved in ways that encouraged local people to keep agreements. The point that needs to be kept in mind since the endorsement of FPIC by community groups is that the whole process of agreement-making reflects the state and the challenge of community relations. Experience suggests that agreements, partnerships, and understandings can only be expected to add value if good community relations are already in place.

Where trust funds have been used for disbursement in Australia there have been frustratingly slow rates of disbursement, to which lawyers and accountants, having limited understanding that overseas aid works in a much less formal way, have contributed significantly. Some familiarity with international development assistance as well as agility and simplicity can produce a better result. Once complicated modern systems for planning, budgeting, and the hiring of staff are put in place they acquire a life of their own.

According to anthropologist Bob Tonkinson, there seemed to be a feeling that if Aboriginal people wanted to participate in the modern economy then they had to start their relationships with outside authority by demonstrating the competence that the process of contact was itself supposed to produce.[21]

## Social Performance

The idea of "social performance" is sometimes mentioned by mining companies in connection with their position on CSR and seems to indicate that there is some sort of public display to be seen, watched, recorded, and given credit. When mining companies talk about their "social performance" it is interesting to wonder what would not be considered to be "social performance." Walking your dog in a village full of dog lovers can be seen as a contribution to society, as is cutting the lawn to avoid upsetting neighbors with tidy properties. To avoid being interpreted as public relations "social performance" would need to be employed in ways that show the social nature of what has been done, in other words, the actual fit with a real society and its beliefs, values, and attitudes. When "social performance" is used by a local company, citizens and critics will be able to supply the context that is needed to decide whether or not the statement is true. Beyond sounding a little pretentious, what would be the point of a local company talking about social performance? Very little. Moreover, the level of opacity is increased when "social performance" is used by a company to characterize the impact of its worldwide activities at international, national, regional, and community level. Obviously, we need to be able see evidence that this is not an attempt by the company to differentiate itself from its competitors and gain competitive advantage by simply using suggestive language.[22]

## NOTES

1. See Peterson (1993).
2. "Corporate social responsibility: making good business sense," see Holme and Watts (2000), Elkington (1997), McIntosh et al. (1998), Zadek (2001).
3. On the evolution of the corporation see Maitland (1908).
4. See Holme and Watts (2000).
5. See Knight (1941) and Herskovits (1941).
6. See UN Department of Economic and social Affairs (2005).
7. In the Tipperary Hills section of Syracuse, New York, where enthusiasm for Irish republicanism is strong there is an upside-down traffic light, and it is upside down because the residents never want the red to shine above the green.
8. The listen, learn, and then engage advice was introduced to Rio Tinto by my colleague Tom Burke CBE.
9. See O'Faircheallaigh (2006).
10. See Macintyre (2007).
11. See UN Department of Economic and Social Affairs (2005), Morrison (2014), Metx (2006).
12. See Buchanan and Tullock (1962), Gauthier (1986), Hobbes (1994), Rawls (1999), Locke (1960).

13. The extraordinary progress of physics, mechanics, and mathematics during the eighteenth century involving Newton, Galileo, Copernicus, Descartes, Leibnitz, Huygens, Kepler, Francis Bacon, and R. Boyle produced an extraordinary effort to interpret social phenomena, in the same way that mechanics was thought to have so successfully interpreted physical phenomena. Hobbes and others began to study man as a physicist studies physical phenomena, rationally but objectively. Hobbes compared death with the stopping of a watch mechanism. The human soul is interpreted as a movement as regular as any motion studied in mechanics. "Vita motus est perpetuus," says Hobbes. Hobbes, *Leviathan*, Introduction, Opera, 1.
14. See Evans-Pritchard (1965).
15. See Chambers (1981). See also on long-term fieldwork, Shokeid (2007).
16. Two views of impact. At Bentota beach, Sri Lanka, marine biologists were concerned by the negative impact they said was being caused by local people diving down to the coral reefs to bring up more and more coral which they turned into lime. The divers were delighted by their increased incomes that had been made possible by being able to get better diving equipment from tourists. They said they had always dived and that the coral would regenerate.
17. See Williams (1939).
18. See Franks et al. (2009).
19. See Kemp and Owen (2013).
20. See Joyce and Thompson (2000).
21. See Tonkinson (2007).
22. See Demtchev (2004).

## Bibliography

Buchanan, James, and Gordon Tullock. 1962. *The Calculus of Consent*. Ann Arbor: University of Michigan Press.

Chambers, Robert. 1981. Rapid Rural Appraisal: Rationale and Repertoire. *Public Administration and Development* 1: 95–106.

Demtchev, Nikolay. 2004. Corporate Social Performance as a Business Strategy. *Journal of Business Ethics* 55(4): 395–410.

Elkington, J. 1997. *Cannibals with Forks: The Triple Bottom Line of Twenty-first Century Business*. Oxford: Capston.

Evans-Pritchard, E.E. 1965. *Theories of Primitive Religion*, 24. Oxford: The Clarendon Press.

Franks, Daniel, Courtney Fidler, David Brereton, Frank Vanclay, and Phil Clark. 2009. *Leading Practice Strategies for Addressing the Social Impacts of Resource*

*Developments.* Brisbane: Center for Social Responsibility in Mining, Sustainable Minerals Institute, The University of Queensland & Department of Employment, Economic Development and Innovation, Queensland Government.

Gauthier, David. 1986. *Morals by Agreement.* Oxford: The Clarendon Press.

Herskovits, Melville. 1941. Economics and Anthropology, A Rejoinder. *Journal of Political Economy* XLIX(2): 269–278.

Hobbes, Thomas. 1994. *Leviathan,* ed. Edward Curely. Indianapolis: Hackett.

Holme, Richard, and Phil Watts. 2000. *Corporate Social Responsibility: Making Good Business Sense.* Geneva: World Business Council for Sustainable Development.

Joyce, Susan, and Ian Thompson. 2000, February. Earning a Social License to Operate: Social Acceptability and Resource Development in Latin America. *The Canadian Mining and Metallurgical Bulletin* 93(1037): 49–53.

Kemp, D., and J.R. Owen. 2013. Community Relations and Mining: Core to Business but not 'Core' Business'. *Resources Policy* 38(4): 523–531.

Knight, F.H. 1941. Anthropology and Economics. *Journal of Political Economy* XLIX(2): 247–268.

Locke, John. 1960. *Two Treatises of Government,* ed. Peter Laslett. Cambridge: Cambridge University Press.

Macintyre, Martha. 2007. Informed Consent and Mining Projects: A View from Papua New Guinea. *Pacific Affairs* 80(1, Spring): 49–65.

Maitland, F.W. 1908. *The Constitutional History of England: A Course of Lectures.* Cambridge: Cambridge University Press.

McIntosh, Malcolm, Deborah Leipziger, Keith Jones, and Gill Coleman. 1998. *Successful Strategies for Responsible Companies.* London: Financial Times Management.

Metx, Shannah. 2006. Indigenous People's Right to Free Prior Informed Consent (FPIC) and Project Governance. *Collaborator for Research on Global Projects.* Stanford University.

Morrison, John. 2014. *The Social License,* 75–79. London: Palgrave-Macmillan.

O'Faircheallaigh, C. 2006. Mining Agreements and Aboriginal Economic Development in Australia and Canada. *Journal of Aboriginal Economic Development* 5(1): 74–91.

Peterson, N. 1993. Demand Sharing: Reciprocity and Pressure for Generosity Among Foragers. *American Anthropologist* 95(4): 860–874.

Prno, Jason. 2013. An Analysis of Factors Leading to the Establishment of a Social License to Operate in the Mining Industry. *Resources Policy* 38(4): 577–590.

Rawls, John. 1999. *A Theory of Justice.* Cambridge, MA: Belknap Press of Harvard University.

Shokeid, Moshe. 2007. From the Tikopia to Polymorphous Engagements: Ethnographic Writing Under Changing Fieldwork Circumstances. *Social Anthropology/Anthropologie Sociale* 15(3): 305–319.

Tonkinson, Robert. 2007. Aboriginal 'Difference' and 'Autonomy' Then and Now: Four Decades of Change in a Western Desert Society. *Anthropological Forum* 17(1): 41–60.

UN Department of Economic and Social Affairs. 2005. *An Overview of the Principle of Free, Prior and Informed Consent and Indigenous Peoples in International and Domestic Law and Practices.* New York: UN Department of Economic and Social Affairs, Division for Social Policy and Development Secretariat of the Permanent Forum on Indigenous Issues.

Williams, F.E. 1939. *The Creed of a Government Anthropologist.* Sydney, Australia: Presidential Address Australian and New Zealand Association for the Advancement of Science.

Zadek, Simon. 2001. *The Civil Corporation: The New Economy of Corporate Citizenship.* London: Earthscan.

CHAPTER 9

# Resettlement

Because of numerous accusations of CSR failure resettlement has been an important topic for mining companies. A number of development assistance agencies have produced standards for resettlement; it almost seems that they are in competition with each other since the IFC has one set of standards, and the World Bank another. As with CSR the standards do not require a good baseline, experience with community engagement, or building hands-on skills. No systematic attempt to grasp the social factors that dominate the lives of those to be moved and which they would like to be maintained is called for. One might assume that the most important party in a resettlement would be the community that is to be moved but this is not the case: local people are sidelined because resettlement has become a matter between the owner of the resettlement standard and the mining company.

Resettlement using International Standards is all about tick-box responses and the middle-class materialistic values of the industrialized countries. Instead of a concern to preserve religion and belief systems there is a concern for incomes, like-for-like replacement, and individual entitlements. The aid agency tick-box one-size-fits-all aggregation of rights, duties, and entitlements related to resettlement apparently has no need of relationships or of becoming a community. There is no mention of leadership, social solidarity, or the ability to impose and accept discipline for the good of all during times of stress and strain during transition.

© The Author(s) 2017
G. Cochrane, *Anthropology in the Mining Industry*,
DOI 10.1007/978-3-319-50310-3_9

Understanding resettlement requires social imagination because this process should involve an attempt to move history and memory, social organization, ways of behaving and getting a living, beliefs, imagination, and creativity. The International Standards provide little help or guidance about the lethargy, apathy, and depression which are commonly seen in resettled communities. Instead there is an emphasis on individuals which can result in too little attention being paid to maintaining, sustaining, and enhancing social solidarity. It is important to recognize that the community that is moved, and not International Standards and the agencies that enforce those standards, is the ultimate judge of how well the move has been carried out.

Resettlement needs community relations that include a social baseline, an understanding of how the community works, consultation, and local development ideas. Above all resettlement needs continuing social relationships between those who are to be moved and those who will do the moving. For many resettlements impoverishment in an economic sense rather than wealth creation has been the result. There is no alchemy that enables those following International Standards to take people with low levels of skill and a low level of enthusiasm for hard work and transform them into productive farmers or traders. The fact that such a radical change is frequently assumed by the International Standards guidance and livelihoods enthusiasts should be continually challenged and corrected.[1] If industrialized countries' governments had the capacity to produce wealth creation among poor people by relying on their policies, procedures, and standards their big welfare bills would disappear.

International Standards suggestion that there can be like-for-like swaps of bits of the environment in the resettlement process cannot be expected to work when the area to be given up is invested with ritual and social significance acquired over many years. When it comes to an environment, a landscape with a history and a place in the belief system there is no like-for-like swap. If it is a territory then that territory needs to be understood in terms of meanings and memories that encompass sights, sounds, and smells, birds, animals, trees, and plants, and so on, that for those born and raised in it is like no other.

Although Rio Tinto undertook the resettlement of communities in a number of countries in conformity with International Resettlement Standards, it was plain that these were not sufficiently focused on the realities of resettlement to do the job. Satisfying the lists of dos and don'ts seems to be more important to the aid agencies than satisfying the community being moved. The International Standards for resettlement spend far too much time on what should happen before a move and far too little on what is needed after a move. It simply is not possible to recompose community life on the basis of one of the Resettlement Standards that are now available. In their defense, however, it is perhaps doubtful that they were designed to be more than rather general beginning guidance. They are an accounting list, and, while they can serve to remind us about important social factors, they are too sketchy and brief to provide the necessary blueprint. Lists of dos and don'ts have their uses but there is always a danger that something important will be left out, and something that cannot be made to fit in will be included in the advice.

Listening, learning, and then engaging after doing a baseline and building relationships based on trust and mutual understanding will go a long way toward achieving success. However, for some unfathomable reason the resettlement of communities is often discussed as a special kind of issue that requires a particular administrative technology not found in community relations. This is unfortunate because resettlement badly needs to rely on the relationships between those who help and those being moved in line with the approach used in Rio Tinto community relations. Rio Tinto experience suggests the need for a 10–29-year plan and for closer attention to be paid to the fact that companies have to move social organization as well as individuals who have individual monetary and other entitlements.

What settlers need is a group of familiar helpers who they know and who speak their language, who will be with them every day explaining, supporting, and encouraging. Resettlement may need to be worked out and changed and altered on a daily basis in the light of developments. It is thus vital to have community personnel in place who are known and trusted. It has always seemed to the author that the most important thing is to be there on the ground with the settlers giving a helping hand.

What is needed is the unscripted and unprompted performance of community workers who know enough about local society and have the personal skills to create and sustain daily contact with a wide variety of individuals. Some social niceties can be passed on to a man or woman making calls each day to unfamiliar householders in order to sell double glazing or a new washing powder but these techniques will not help someone who has to behave like a community resident. A welder welds to order, a mechanic fixes according to the manual, a teacher teaches according to the syllabus, a health worker heals according to the diagnosis, but a community worker dealing with resettlement has to come up with something new over and over again in order to be influential. If the miners in the community are not up to the task, NGOs and aid agencies will urge that priority be given to their list of things that must be done and will also urge acceptance of the views of their staff even though they may not know the country, the community, or much about community relations.

It is not hard to see that both the company and the community to be displaced need to be satisfied that this move is really necessary. After all, in some instances the company may want to move people out of the way simply to allow mining to begin. In other circumstances population increase or a deteriorating resource base may have led a community to want to move. While this decision is important and while it must be documented, what is also necessary is to begin forward thinking: to sketch out what should happen over the next 15–20 years.

What is surprising given the fact that aid agencies, though responsible for providing advice on economic development, have not included in their lists of what should be done any requirement that those resettled should receive some portion of the wealth created by the projects necessitating resettlement. Without resettlement that wealth creation would not have been possible. What often seems to happen where aid agencies are involved is that the resettlement is wrapped up in one project whereas the wealth creation occurs in another without any obvious linkage. Surely, in relation to resettlement for dam construction, for example, the community involved should not only receive free electricity but also some share of the profits?

## PHOENIX ISLANDERS

If you look very, very hard indeed at a map of the central Pacific Ocean you may find the Phoenix Islands, which were part of the British Gilbert and Ellice Island Colony. For 18 months in the 1960s I was responsible for a resettlement of 1100 Phoenix Islanders many of whom had worked mining phosphate on Ocean Island.[2] The population of the Phoenix Islands had been about 1100 people who lived in three communities: Orono, Nikumaroro, and Kukutin. Two main religious groups were represented: the London Missionary Society and the Roman Catholics. After the three communities were moved to their new home the resettlement had to be organized on a community basis because, owing to religious differences, the two communities would not work together. Even the dancing was different, because of differences between the Catholics and the London

Missionary Society. Catholics danced bare-breasted with a movement of the hips sometimes known as a washing machine action, whereas London Missionary Society dancing was much more restrained.

The Phoenix Islanders had to move because of a drought lasting almost seven years. Their coconut trees were dying, wells produced only brackish water, and, indeed to outsiders, one of the remarkable things about the young children was that they had never tasted fresh water. The authorities were carting water by a slow-moving ship to the Phoenix group from Tarawa, thousands of miles away. The Phoenix Islands consisted of low coral atolls; it was said the islanders would get vertigo if they got 15 feet above the ground. The agriculture was hard; pits were dug and filled with compost and then a few vegetables could be grown. Fishing was a major activity. Money had been earned from copra production and also from working heavy machinery in the phosphate industry on Ocean Island. Others had worked on Christmas Island when Britain was carrying out atomic testing.

The government established a resettlement team to do the job. Each member of this resettlement team spoke Gilbertese. The team contained marine, health, and agricultural personnel, and its first task was to visit the Phoenix group to obtain the permission of the islanders for a move. After this was done, the team began to survey the existing way of life of the Phoenix Islanders as a prelude to designing an appropriate reception for them in the Solomon Islands.

The settlers, and their canoes, arrived by ship at Wagina Island in the Manning Straits of the western Solomon Islands. Immediately the resettlement team faced several unforeseen challenges. When the settlers first got in their canoes to go fishing they were unfamiliar with the strong local currents and several had to be rescued while on what appeared to be a course for Japan. To counter the fact that fishing by canoe would not produce the amount of food that had been planned the team arranged for a daily catch of fish by trolling and netting them using project boats. This produced around 2000 pounds a day. Gilbertese were fond of dried shark and the liver was particularly nutritious. Cookies, sugar, and tea were also provided for each family. There was concern that putting people on rations would lead to difficult times when rations had to be withdrawn and it was already uncomfortably obvious that the per capita cost of rations was almost the same as the cost of feeding prisoners in a government jail.

Malaria had to be to be controlled. How did one ensure that 1100 people who were not used to popping pills engaged in regular prophylaxis?

Mosquito nets were issued but they were soon turned into ladies underwear. Visitors arrived by canoe (at odd times and in odd places on Wagina) from Bougainville or Choiseul and many were malaria carriers. The settlers did fine and the author was the only person who caught malaria. One or two people did swallow quite a few anti-malarial pills at a time but it did not seem to do them any lasting harm. We built latrines over the reef and covered the sides of the structures with coconut fronds. The Gilbertese removed and re-used all the fronds and resisted replacements since while they performed their bodily functions they valued conversation with those on the platform or on shore more than privacy.

Settlers began to build houses but both house construction in the Solomon Islands and the materials used were not familiar to the Gilbertese. We arranged for a team of Solomon Islanders to come from Choiseul to show the settlers how to make houses the local way, using local materials. In the Gilberts coconut fronds were used to roof houses, but these roofs only lasted for three years or so. Solomon Islanders used nipa palm rather than coconut, and this lasted for five or six years. The timber that was used for house poles in the Solomon Islands was different and so was the trunk of the betel nut tree, which was used for flooring. Nails were used in house construction by Solomon Islanders rather than the sinnet cord made from coconut fiber that was used in the Phoenix Islands. The Gilbertese were very slow and methodical. The floors had to be absolutely level. The house posts had to be well seated. The roof had to be carefully put on. The result was that while we had estimated two months for house construction the task took almost three times as long.

Around 400 acres were made available to the settlers, and when added to the other suitable ground, this meant that each family was to be allotted around 10 acres. Settlers had to learn about new crops. The agricultural objective of giving the settlers a good cash income had to be achieved as soon as possible, because government did not want the settlers to be on rations for too long. However, it was going to take five to seven years for the tree crops to produce income-earning crops. Also, before the planting could take place, heavy stands of tropical rain forest had to be cleared. Never having cleared such large trees before, the Gilbertese found tree-felling difficult. So did onlookers, since at first the felling accuracy of the settlers was in the 360 degree range.

The medical officer, a doctor who had trained in London, had the habit of looking in his books and then declaring that nursing mothers needed this vitamin or that food supplement intake. He knew next to nothing about the value of indigenous foods such as shellfish, coconut, and wild vegetables from the forest. The most serious and embarrassing problem, since the Phoenix Islanders had been moved to avoid drought, was that the resettlement area was short of water. A former marine officer (Tekinaiti Tokatake) and an old man called Tabea suggested building wells because Gilbertese were used to wells. Several were built and that solved the water problem.

Although the settlers had been moved because of a lack of water in the Phoenix Islands the situation was not much better in their new location. The government had relied on 1947 US Air Force aerial reconnaissance photographs, which turned out to be inaccurate. The settlers who had been miners on Ocean Island came up with the solution themselves. They suggested that wells be constructed with concrete shuttering. Gilbertese were used to wells and to the sort of behavior that well use required in order to avoid pollution.

## Zimbabwe

Rio Zimbabwe (RioZim), a subsidiary of Rio Tinto, was in the middle of the process of planning and getting approval of a resettlement from Robert Mugabe's government when an official announced: "We do not like your gold-plated resettlement." RioZim company needed to move more than 100 families who were living on top of a diamond deposit to a new location. The company had commenced community consultation and engagement early in the project and by 1999 discovered that the mining lease would require 1200 hectares of land, which was occupied by 142 families needing to be resettled in order for mining to proceed.[3] Resettlement was the major issue since the mining was thought to be relatively simple. One executive joked that his mother-in-law could do it with a wheelbarrow. At the same time 265 graves needed to be relocated before mine construction commenced.[4] Thirty families decided that they did not want to move, which was worked around though there was concern as to whether these families would be able to gather firewood after the move and whether fencing around the mine and the construction of a weir on the river might affect traditional patterns of movement.

There never was any formal legal agreement. The relationship-building that the company had engaged in over a number of years had convinced people that the company would keep its word. RioZim began by offering each family 80 hectares which would be fenced and stumped to remove trees that would impede tilling the soil. At this point the Mugabe government announced that they were sending 200–300 "War Veterans" to the area to be resettled. This meant that the company would have to accommodate the settlers who were from urban areas in Masavingo, which was reputed to have the highest incidence of HIV/AIDS in Zimbabwe. What was even more problematic was the fact that the "War Veterans" had little background in agriculture. To accommodate the newcomers RioZim was forced to halve the amount of ground the Murowa settlers would receive, though this reduction was softened slightly by promising to provide land for common grazing.

The resettlement exercise began with a sort of Domesday Book exercise at Murowa which involved the resettlement team making an inventory of all that the settler families possessed and getting from them the design of the house, kitchen, and storage facilities that they would receive when they had been resettled. The original survey was signed by the team and the householder and after resettlement had taken place and all snag lists had been cleared both parties then signed the document to show that promises had been kept. Arrangements were made to have Murowa people go to the new resettlement site so that they could think about planning and become familiar with the area. It was also important that those who had crops in the ground at their old location could harvest them and know that they also had food in the ground at their new homes.

The diamond mine was located near Zvishavane in southwest Zimbabwe. Exploration began in the region in 1992, and subsequently a small-scale operation with potential to expand should the current local socioeconomic, political, and world market conditions improve to allow for further investment was developed. The first diamonds were produced in 2004, and the operation employed around 100 permanent staff and 200 people on long-term contracts.

It was not a good time to be looking for land because expatriate farmers were being expelled and their lands taken over by the government. The company managed to buy six blocks of land at a place called Shashe but the legal aspects of the purchase were worrying. The white owners might later complain that they sold under duress. The infrastructure agreement that was negotiated with the local authorities included provisions for RioZim to build a school, rural health center, housing for teachers and nurses, a church, 17 boreholes, and roads, as well as setting a timeline for their construction and eventual handover to the government. Meanwhile the allocated land underwent improvement and preparation, allowing the resettled communities to plant a crop immediately, and reap a valuable harvest in the 2002–2003 season.

At the handover of the Shashe resettlement scheme, the company publicly committed itself to a Community Action Plan that provided for health, training, and agricultural capacity-building programs for 10 years after resettlement. The aim of the program was to offer agricultural skills and capacity building to enable the farmers to run farming activities that could sustain them. Other initiatives included running a national training program for farmers and introducing hybrid cattle and high-yield crops.

"Social jealousy" was a major concern given that RioZim's settlers were due to receive more assistance than the "War Veterans." How would the "War Veterans" react to the fact that they had very little help, while their company-moved neighbors had their land stumped and plowed? Boreholes had been sunk, roads constructed, good houses, water stand-points, kitchens, and granaries provided. The other settlers had not had anything because the authorities had very limited resources. The settlers voted in favor of being integrated with the others. In practical terms help cannot be easily delivered unless they are all treated as one. Attempts were made to include the War Veterans which helped to ensure that relation-ships between the two communities remained stable. It was not easy in a country that had been dominated by big farms and farming-systems thinking and practices to get people who deal with subsistence. It was all very well to plow and plant but experience suggested that the ground would turn into concrete without irrigation and this would require con-siderable investment. Off-farm employment would be very important in future years.

In the end what was called the diamond resettlement was, in fact, three resettlements, involving (1) those who were left behind and whose lives had to be adjusted to the loss of a significant number of community mem-bers, (2) those who left the mine and went to the new area, and (3) the resettlement accommodation of the "War Veterans."

## GUINEA

Rather than the mechanical tick-box application of International Standards, Guinea represented an example where traditional management systems and social organization could prove to be useful. Guinea resettle-ment planning, which is still ongoing, considered the possibility that key roles might be played by the elders and the traditional segmentary lineage system. Resettlement would be necessary to facilitate the construction of a 750-kilometer railway from a mine in the east of the country to a port on the western coast. However, the timetabling of this iron ore project has been affected by the outbreak of Ebola. That has provided more time to address some of the complexities which include that fact that along the rail corridor there are strikingly different agricultural production and livelihood strategies, ranging from cash cropping to cattle rearing and traditional subsistence forest activities. Six distinctively different ethnic groups would be affected, some of which had experienced relatively little contact with the outside world. The majority of Guinea's population is rural, and more than 70 percent of the population worked in the agriculture, livestock, fishery, forestry, and mining sectors. Farms are family-owned and operated, and generally small: two-thirds are less than three hectares. Pastoralists move large herds seasonally between the hinterland and the coast, negotiating with the settled farming communities for access to dry-season grazing and saltlicks in the grasslands and coastal plains.

Using the results of the social baseline data collection, two methods of land acquisition and resettlement were developed for testing. The first, aimed at those areas where cash cropping and individualization of property concepts was well entrenched, could use international standards resettlement and land registration as a way of providing security of tenure to local people in the more economically advanced areas at either end of the rail corridor. Many of Guinea's formal land laws and policies recognize customary land rights but there remains a distinct gulf between statutory policies and customary practices. Under the Land Code, rights must be registered, but state land administration institutions lack capacity and resources to support registration or have never been created. The 1992 Land Code was formal and uniform in nature but had little recognition of the existence of community-based tenure systems.

The second method would recognize the unacceptability of land registration in remote rural areas and would take into account the critical role of the lineage in traditional rural areas and seek to facilitate a process whereby the lineage would allocate land to those who had to be resettled. The second method took into account the fact that land along the corridor in traditional areas was typically communally owned and managed by lineage members for the greater good of a larger collective. Traditionally, these lineages have a responsibility to find sufficient land for their members. The trend in West Africa was to move away from national, homogeneous land codes that did little to recognize local practices. The new trend in tenure policy was to explore community-based solutions to tenure insecurity and a state-facilitated evolution, rather than replacement, of indigenous land tenure systems. The neighboring countries of Mali and Burkina Faso had moved to more decentralized tenure policies that place a renewed emphasis on community-based tenure rules. Continuing specialist review is required to assess the fit that can be achieved by legal innovations from other countries given the situation in rural Guinea.

Would this lineage process provide for security of tenure for those being moved? Would titles have any value? While this would ultimately be an empirical question, there was good reason to believe that an out-of-date or fairly weak title still confers considerable protection. Firstly, in most cases of inheritance the title would bear the same family name as that of the current owner. Secondly, the issuance of the title, even if many years in the past, would imply that the parcel is part of the title deed registry and its boundaries and title number that appeared in the cadastral record at the office of land administration. Consequently, it would be extremely difficult to have a new title issued for land incorporating a previously titled parcel, even one subsequently subdivided among several co-inheritors. Certainly, it would be far easier to exploit the modern titling system to nullify an informal ownership claim than a formal one.

Moving the elders would provide reassurance to settlers that traditional forms of land management would continue in their new location. This has meant recognizing the role and authority of the village elders and encouraging their assistance in the resettlement process. The elders manage village lands; they allocate land and ensure that it is used for the benefit of all. An underlying principle of customary land tenure systems is that within a community the lineage elders will usually try to make sure that everyone has access to the agricultural lands required to meet his or her needs. The lineage performs a valuable land management function by making sure that the various types of land held by the lineage are allocated to meet the needs of a growing population. It also makes sure that good use is made of these common resources. In order to ensure that all members of the lineage have the land resources they need, the area of land cultivated by each person may vary from year to year, depending on the needs of members of the lineage. An exception is upland areas used for tree plantations. In this case, the person responsible for the plantation is normally granted consolidated rights to the land which he can leave to his children as part of his inheritance. As a consequence, he must have the approval of the elder of the lineage prior to establishing his plantation.

The role and functioning of the elders would need close examination and continuing close contact. Of course care would have to be exercised because in some areas young men and women might not be keen to have a continuation of traditional authority. The elders are a grass-roots institution and in effect have to negotiate their power day by day and, therefore, embody a degree of flexibility that is extremely useful for the efficient management of natural resources. The physical closeness to their constituency allows for the application of a set of rules and norms that will rarely be out of touch with the ecological reality and the management and conservation requirements of the resources in their territory.

The head of the lineage was generally a descendant of the first occupant of an area and, therefore, the guarantor that the original pact with spiritual and terrestrial powers of the earth would be respected. Local names for the *chef de terre* translate literally as "master of the land," "child of the land," or even "owner of the land." The land chief holds substantial powers over land in customary societies and is an institution which is both sociopolitical and socioreligious.

Traditional chiefs acted as mediators between a given ethnic group and its environment, and their relationship to the land is determined by the location of the burial places of the ancestors of a given lineage (or tribe). Traditional community leaders are the symbol of an intimate alliance with their territory.[5] The primary function of such traditional authorities is to ensure peace and harmony in the rural communities within their territory. Thus, a bad chief or sheikh would be someone not able to assure this, for example, during celebrations when people consume alcohol and fights may break out. The main tasks of traditional authority revolve around mediating in land-conflict resolution and regulating access to land. Solutions are normally reached among the parties involved, often with the mediation of the respective local lineage chief(s) or sheikh. Only when the latter are unable to reach a verdict acceptable to everyone is the traditional chief or the grand sheikh (sheikh daman) approached. He or she is acknowledged to have ultimate knowledge of the customary geographical boundaries and will take a decision, in consultation with his or her counselors. In performing these services, traditional leaders have obligations and duties without any rights. There is no reward for the work they carry out for the institutions of the government, or for organizations or companies, which occupies their time to the detriment of their fields.

## Lineage Resettlement[6]

Was resettlement on lineage lands possible? Local people could be expected to want to use their lineage affiliations in order to seek adoption by lineages of their kindred in nearby communities and hamlets that had available land. Surveys suggested that there might be sufficient lineage land available to be able to provide for kin who were to be resettled. Could those who were members of a lineage (and who had to be moved in the traditional non-monetized areas along the corridor) be moved to other lands held by their lineage kin? Obviously, local people were unlikely to want to resettle on land where they had no pre-existing lineage affiliation. Surveys were needed to understand the extent of the lineage system role in Guinea; the extent to which communities, villages, and markets were linked by kinship; and the extent to which the lineage was relevant for contemporary village life. Kinship, specifically segmentary lineage systems, has provided the mainframe for village life in Guinea and other African communities. Kinship can be the government, the legal system, the political system, the economic system, and the overall regulator of village life. In the traditional system there are two major categories of rights: collective rights of the lineage and its members, and individual use rights possessed by members of the lineage.

## Notes

1. A World Bank/ADB resettlement in Lao PDR spent a great deal of money per settler and had all kinds of technical advice but did not demonstrate that the World Bank could itself produce a best-in-class resettlement. "Nam Theun2 World Bank/ADB Resettlement," International Rivers Network, Berkeley CA, 2007.
2. Although I had been instructed in the elements of the Gilbertese language in London before going to the Solomon Islands my skills were weak. I received an enormous amount of advice and help from William T. Stuart, a young anthropologist who was studying the resettlement. See Cochrane (1969).
3. I worked with Dr. Carolyn McCommon, Frank Webster of RioZim, and Paul Bundick of USAID on this project for a number of years.
4. The Plateau Tonga were resettled as a result of the construction of the Kariba Dam in the 1950s, and they are still discussed in terms of having been resettled. The resettlement was very poorly planned and executed. See Scudder (2007).

5. *Mapping traditional structures in decentralisation policies* [online] Available at: http://www.fao.org/docrep/006/ad721e/ad721e03.htm-fn21#fn21 [Accessed 21 September 2016].
6. The idea of Lineage resettlement was based on a process that seemed to occur naturally though it was also obvious that this process had been made more difficult as a result of the arrival of refugees from conflict in Sierra Leone and Liberia. On the way these lineages operate see Bohannon (1957), Evans-Pritchard (1940).

## BIBLIOGRAPHY

Bohannon, Paul. 1957. *Justice and Judgement Among the Tiv*. Oxford: Oxford University Press.

Cochrane, Glynn. 1969. The Administration of Wagina Resettlement Scheme. *Human Organization* 29(Summer): 123–132.

Evans-Pritchard, E.E. 1940. *The Nuer: A Description of the Modes of Livelihood and Political Institutions of a Nilotic People*. Oxford: The Clarendon Press.

Scudder, Thayer. 2007. Pipe Dreams: Can the Zambezi River Supply the Region's Water Needs? *Cultural Survival Quarterly* 31: 7–31.

# CHAPTER 10

# Results

Mining companies looked to UN agencies because they had implied that they knew what was best for communities that they did not in fact know very well at all. Rio Tinto and other mining companies were mistaken when they assumed that UN agencies and big NGOs had the local skills and hands-on community experience needed to judge, supervise, and improve the performance of miners in the community. Too few of the UN staff, who had been hired from elite universities, had substantive knowledge of and experience with agriculture or health or even of living and working overseas for lengthy periods. Even fewer had ever met and worked with the very poor people whose lot their employing organizations said they would improve. And if the new entrants had not acquired in-depth community knowledge and hands-on skills before joining their agencies it was unlikely that this deficiency would be remedied by their UN service in Washington, New York, or Geneva. Today the UN and big NGO staff spend their working lives in cities in the industrialized countries: 10,000 of 13,000 World Bank staff live in Washington, DC, and UN staff live and work in New York and Geneva when they are not making short trips overseas.

In pursuit of their priorities UN agencies and, increasingly, NGOs have become publishing houses showcasing their aptitude and willingness, if funded, to provide leadership in response to every problem faced by the planet. The World Bank produced a World Development Report every year as well as research papers on every imaginable developing country choice or issue.[1] UNDP produced its Human Development Report, a

© The Author(s) 2017

G. Cochrane, *Anthropology in the Mining Industry*,

DOI 10.1007/978-3-319-50310-3_10

document which tried to go one better than the World Bank by putting in tables on just about everything that could conceivably be related to poverty and the development process.[2]

Unbridled global ambition resulted in hands-on community skills being squeezed out of international development agencies, and big NGOs, like toothpaste from a tube. For the UN agencies and big NGOs the development of presentation skills has become an imperative because of the need to raise more and more money and the need to be seen to have an increasing market share of the aid business. Global aid organizations wanted staff that could do well on paper in order to sell projects, programs, and intellectual products to developing country governments and private sector developers.[3]

It is a pity, given the very large number of UN organizations that exist, that there is no agency that has a remit to work on the ground at local government level. Working from offices far away it is very hard indeed for UN staff to reach communities or poor people because an organization with a local focus is required.[4] Organizations with a global remit spend most of their time and resources (perhaps 80 percent) explaining, promoting, and negotiating the adoption of their technological and intellectual products with senior people at the top of governments and in international public, private, and civil society organizations. Much of the advice relates to the management of the economy and the environment, and social and community matters are usually not included. After obtaining high-level endorsement, adoption and implementation is regarded as a government- or user-organization responsibility requiring only a small portion (say 20 percent) of the organization's time, resources, and skill base for lower-level persuasion and extension.

Organizations that see themselves as having a local focus approach their mission in a very different way from global organizations. Take, for example, Australian organizations with a national remit for rabbit and weed control. They need to secure coordinated action on the ground by farmers, mining companies with operations in rural areas, and communities.[5] If these locally focused organizations do not succeed in getting their message across family, industry, and community boundaries they cannot succeed because isolated control has limited impact. Technological control solutions are available but those responsible for rabbit and weed control have concluded that only a small part of successful implementation (about 20 percent) of the problem is technical while the remainder is social. The implication of this is clear: if organizations that have a national remit to control the rabbit and weed problem in Australia do not develop the hands-on skills and local knowledge sufficient to persuade communities to cooperate, their control effort will fail.

In the early days of the World Bank it was not unusual to find staff called "Loan Officers." These men and women who had practical hands-on community experience acquired during their service in their country's colonial service ensured that specialist staff paid attention to local factors and they were very good at establishing relationships with a wide variety of people in poor countries.[6] British loan officers were working for the World Bank on former British colonies, the French working on former French West Africa, the Dutch on Indonesia.[7] But by the early 1990s loan officers had disappeared.[8]

Community relations will probably continue to be an orphan in the UN system because the behaviors associated with presentation skills and those associated with community skills do not thrive in the same organization. Those with hands-on community skills and those with presentation skills are different people with different priorities and ambitions. These differences usually result in fractious relationships, in much the same way as planners and budget experts do not get on well together in aid agencies.

What to do with and about UN agencies deserves a somewhat more considered review. However, the mining industry association ICMM has perhaps been too close and too involved with international agencies such as the IFC to do what was needed to challenge the waste of time and money involved in the international reporting regime or the weaknesses of the aid agencies and NGOs at community level. Confidence in ICMM's grasp of these issues is unlikely to be boosted by that organization's recent release of a communities tool kit.

## COMPLIANCE

The complicity charges in the 1990s had convinced Rio Tinto and other mining companies that they needed to improve performance in the community so they made systematic attempts to develop hands-on capacity. Beginning in 1995, big mining companies began to establish a reputation for community competence; greenfield sites were being developed and mature operations expanded.

Swaddling the body corporate with tick-box human rights assessments of dubious incisiveness, genuflection in the direction of the UN Global Compact and international environmental standards and the provision of punctilious responses to every information request and tick-box questionnaire diverted time and resources from working in the community but did succeed in earning the miners a get-out-of-jail card, albeit in ways that were often tediously burdensome. A company might complete a human rights survey questionnaire from the Danish Institute for Human Rights, the short version of which is a couple of hundred questions, or they could take the long version, which is four or five times longer.

The completed questionnaires have been used to produce league tables that reduce the problems and the progress of companies in poor countries to progress targets. It seemed to be assumed that every company (and every community) in the world should have more or less the same goals and address more or less the same problems. If companies do not see that then they have to be helped to understand the logic behind this approach. As a result of the presumed competence that came with tick-box compliance CEOs received fewer letters of complaint, and annual general meetings of company shareholders were quieter.

The determined effort by Rio Tinto to develop competence in the community soon began to show the potential to significantly lessen delays to projects, while improving the public image of the miners and community acceptability plus offering a high probability of being able to meet investment expectations on time and within budget. The approach to new community relations, which was used at every company operation, paid off and added significant commercial value by cutting the time taken to get mines up and running and by countering criticism of the industry. Rio Tinto had shown that community competence could not be achieved or measured by tick-box reporting. The answer had been to use anthropology and the other social sciences to innovate, develop, and keep up to date, best practice skills and to do the hard yards in the community. Investing in social science especially tailored to community relations had raised the level of competence in the industry. The idea of taking relationships with communities seriously had proved its value. Community relations was becoming a new discipline, an emerging profession, a field of activity, and a little more: it encompassed and contained disciplinary, development assistance, commercial and professional characteristics, but with an emphasis on the practical, combining studying and doing the knowledge.[9]

Rio Tinto had taken on board the lesson from Bougainville that indicated a need for the long-term building and maintenance of high-quality hands-on personal relationships between miners and the community characterized by mutual understanding, mutual trust, and mutual respect. The quality and depth of the relationship between the community and the mining company was what really counted. It mattered little for community decision-making that a company awarded itself high marks for what it had done in the community if this was a glossy one-sided view. Nor did the views of those who believed that they were "interested and affected" parties necessarily carry local weight if they were not known to the community.

Bougainville experience had suggested that it was important to take great care to select the right people for overseas assignments and to ensure that they and their families had adequate cross-cultural training before taking up their posts. Only a few years previously human rights abuses at Kelian showed that this lesson had not been taken on board. Cross-cultural training was later developed but was canceled and unfortunately never made a required part of the formal induction process for overseas assignments.[10]

### *Drawing Down Social Capital*

As the volume of criticism over community relations dropped, senior Rio Tinto executives appeared to have decided to save money by taking their foot off the accelerator. Community relations went on a "care and maintenance" basis. Although what was needed to consolidate the progress was a publicly announced strategy for the future and that community relations be regarded as a core mining discipline, short-term thinking took over. The company retrenched all the specialists who worked in Australia and afterwards the managers of the communities function were not able to attract and retain any suitable replacements. By 2014 Rio Tinto no longer had a single senior social specialist in London or Australia who had lived and worked in developing countries for long periods and was familiar with the procedures and approaches not only of mining companies but also of international aid agencies. Although Rio Tinto assumed that the company could go out and hire replacements when they pleased that was not necessarily the case because, unlike the engineering disciplines which had long experience with mining and fit-for-purpose academic training, social specialists had to be developed in-house and that required a number of years. The social specialists' skills were dependent on industry experience that could not be gained quickly. It was both predictable and inevitable that the loss of these specialists would have a negative impact on performance.

Those cutting costs obviously wondered if it was still necessary to do the hard yards in the community? If critics could be defanged with tick-box reporting from the office, glossy self-congratulation, photos of happy smiling faces in the community, references to close working arrangements with those protecting endangered birds and rare flora, and, better still, close working arrangements with NGOs and UN agencies why worry and spend all that money in the community? Public relations skills could do the job, surely it was obvious that the UN agencies and big NGOs had no capacity to either understand or to check what they were being given? One of Rio Tinto's top executives was said to have declared that since so much social capital had been generated around the world some drawdown could obviously be afforded. It began to look as if Rio Tinto saw good community relations as if it were a commodity, something that could be traded like a carbon credit.

There were no community replacements in the pipeline because Rio Tinto, in common with other mining companies, had done a miserably unimaginative job with the development of the skills needed to do well in the community. There was no career path, it was practically impossible for those working at an overseas site to secure a transfer to another site, and the importance of training was ignored. No university scholarships for practitioners or professional in-service programs such as the Young Professionals program of the World Bank Group had been established. Within the past decade there were no examples of sending men and women who had shown their aptitude and appetite for this kind of work at community level to tertiary institutions for further education.

Obviously the universities have not been able to deliver entry-level skills and are unlikely to be able to do so in future. It would have made some sense if the industry had pooled its resources, perhaps included broader extractive industry, in order to develop and deliver a common form of entry level and other training to beginning personnel. Provision of a common form of training with common qualifications would represent a considerable advance over present arrangements. Here the example of the Merchant Marine might be instructive: provision of a common form of training with common standards and expectations organized around different companies. Mining companies such as Rio Tinto have not yet done enough to develop a research agenda that would make good use of the centers for the study of social issues in mining at Queens University in Canada, the Colorado School of Mines in the USA, and at the Centre for Social Responsibility in Mining located at the University of Queensland in Brisbane, Australia.

Clumsy cost-cutting sent a signal to the best and the brightest graduate social scientists that the mining industry was not necessarily a place where one could spend a career doing work which was appreciated and valued. A substantial part of Rio Tinto's community relations success had depended on a clear and deliberate and announced intention to exceed the expectations and standards of UN agencies, big NGOs, and communities. Even when mining was not commercially viable, as in the case of Panama, Rio Tinto spent money and provided technical assistance to provide the Guaymi community with a helping hand in the form of an access road. When company exploration prospects had ended around the world the communities nearby were left with roads and bridges, operating clinics and schools that were sometimes funded for months and years after the geologists had left. These were the things that made people proud to work for Rio Tinto.

The tradition of providing a helping hand, above and beyond what was required by law or regulation, was, unfortunately, not part of the Bougainville exit. Bougainville needed a gesture of goodwill, a symbolic gesture that, while it did not indicate the acceptance of responsibility to fix all that was wrong, or for all the unhappy events of the past, showed some concern for the fate and the future of those that had been partners, workmates, and friends. The Bougainville exit decision provided an opportunity to show how far Rio Tinto had come since the Civil War.

## RUNNING ON EMPTY

Maintaining a strong focus on community relations was difficult to sustain as Rio Tinto became highly centralized. When downsizing came it was the people at the bottom of the organization who were retrenched not those who had authorized the expansion and wasted company resources. As staff numbers had gone from 200 in the 1990s to around 1000 in 2005 more and more work with community relations implications was being done in the office by corporate personnel who were not necessarily familiar with the communities' function or those who worked at site. More and more senior managers who had never lived and worked abroad in a developing country were put in charge of large overseas investments in countries with complicated community issues. As Rio Tinto had expanded its footprint abroad the company had run out of what former US Ambassador to NATO Harlan Cleveland called Overseasmanship skills.[11] On the retirement of the early mining company pioneers, who had discovered the minerals and built the mines overseas, they were replaced by men and women who had never been through a tough NGO campaign.

The new overseas managers had a strong focus on the short-term results rather than a slow and patient building of local knowledge and personal relationships. They did not start with a sense of inquiry about what was possible and how long it would take to achieve the possible. They were not convinced that what those with local experience saw as important really mattered. Yet they did matter and it was not always helpful to have managers who wanted to keep pulling up the flowers to see if they were growing. Meanwhile those in charge of the communities function began a process of bureaucratizing community relations thinking. They commissioned manuals for community relations, stakeholder and legal relations, cultural heritage, human rights, and gender balance (most should have been written by Rio Tinto personnel but were outsourced). The manuals which were simply tool kits by another name could not be considered to be an adequate substitute for regular contact with specialists because community relations personnel need coaching and coaxing rather than being left to read the manuals.

Rio Tinto's most important community possession was not a cabinet full of manuals and procedures and a cadre of personnel skilled in the filling out of forms but well-selected and well-trained practitioners with the ability to think for themselves. Rio Tinto's success with community relations had relied on the assumption that those working in communities would at times be encountering situations none could visualize in advance, and what they needed most was the incentive and the skills to learn about the unique situation and the confidence to invent a way to deal with it. This emphasis on preparing personnel who worked in the community to handle uncertainty had been supported by the creation of a management climate, supported by the CEO at the time, in which differences of opinion were welcomed; mistakes were seen developmentally, rather than causes for punitive behavior; risk takers were rewarded; and the leader's role was more one of supporter than of enforcer. The reverse point needs to be made that rules should not be prized because they are not particularly functional in an organization that seeks to avoid predictability and conformity.

There was obvious confusion at the top. In 2012 Rio Tinto announced that "All operations have locally appropriate, publicly reported social performance indicators that demonstrate a positive contribution to the economic development of the communities and regions where we work, consistent with the MDGs [Millennium Development Goals]."[12] However, at the time the commitment was made neither Rio Tinto sites nor the headquarters in London had the quantitative skills that would have been required to meet this MDG commitment or for the SDGs which have 17 objectives and 169 indicators.[13] Since the commitment was made there does not seem to have been any public reporting.

Aping UN agencies whose environmentalists used standards, Rio Tinto then adopted a standard for communities. A standard was obviously useful for measuring air and water quality but the utility of a standard in the social sphere is less obvious. Use of a standard was not going to help to reach a deeper level of understanding about relationships, which is the business of community relations. Rio Tinto managers seemed to have lost sight of the importance of recognizing the aspirational nature of their work in the community.

In UN agencies social impact and standards thinking has reduced the role of anthropologists, and other social scientists interested in social relationships, to one of only looking at the constraints on progress—SIA, for example—rather than on also developing ideas about how to produce useful social change. UN and NGO anthropologists and sociologists have been engaged on safeguards work related to gender, indigenous peoples, and human rights. The term "safeguards" was used in the old League of Nations thinking about "Native Peoples" needing the protection of reservations and international organizations to save them from the rigors of modern life. Indigenous peoples now get the indigenous project from UN agencies.

The loss of the company's senior social scientists caused the five-year planning system to be abandoned without any adequate replacement. Regular exchanges between product group executives and social specialists lubricated the relationships between the corporate center and sites that were involved in the five-year planning system for community relations. The five-year planning process underpinned the link between communities' personnel and the corporate headquarters. Cooperation between operations and headquarters was based on regular personal contact generated by the five-year plan, making London aware of the detail of what was, and was not, happening at overseas sites.

Ironically the cost-cutting turned out to be a self-inflicted wound because it produced an increase rather than a decrease in costs since without any specialist expertise Rio Tinto had to spend more money on consultants when trying to get greenfield sites up and running. In the absence of Rio Tinto specialists, Rio Tinto staff in the field without social science credentials and experience were unable to write tight terms of reference for consultants, unable to supervise the work closely, or unable to make good use of the reports that were delivered.[14]

Without experienced advisors Rio Tinto Product Groups would inevitably find it difficult to produce an authoritative board paper that covered the social factors and risks involved in increasing or decreasing investment. In Chile, Mongolia, Madagascar, and Indonesia, countries where investments of billions of dollars had been obligated, Rio Tinto no longer had experts who could anticipate, help to avoid, or ameliorate nasty surprises. Without access to social specialists headquarters executives had no option other than to develop their own ideas about what to do in the community by consulting the Big Five companies. These companies did not have strong community skills or specialized country knowledge.[15] However, the very expensive Big Five firm ideas always found favor with senior executives because they were gift-wrapped: they reduced the most complex problem to wonderfully clear, compelling, and logical courses of action; they provided a symphony in Power Point slides, and graphs studded with milestones, containing little gems about strategy and nuggets of market speak. The Big Five consulting firms had the ability to suggest ways for a business to expand and, when that did not work, they could suggest ways to downsize and, when that did not work, could then, unashamedly, move on to their next assignment.

Lacking access to in-depth country understanding, corporate executives had to rely on working with risk specialists. While collectively the executives and risk specialists might know a lot about engineering, commercial, and geological data, they often knew very little about the communities involved. They could note what they thought were the key social issues on a flip chart and could give the mandatory ranking to each issue indicating their assessment of the seriousness of the risk. Then the risks could be graded and different colors used to add visual impact and precision. The results, while reassuring, could be dangerously misleading and expensive.[16]

The communities' document *The Way We Work* helped Rio Tinto to become a better company and there is more value to be generated by a return to the pursuit of excellence. However, this cannot happen without far-sighted leadership, confidence that those supervising and managing the function have adequate technical comprehension and competence, re-acquisition of specialist expertise, and an insistence on creating, rather than drawing-down, social capital. If the lessons of experience remain overlooked, Rio Tinto community relations will continue to slide backwards toward the position the company was in before the import of Bougainville was understood and taken on board.

\*\*\*\*

\*\*\*\*\*\*\*\*\*\*\*\*\*\*\*\*\*\*\*\*

An apocryphal conversation between an African Rain Doctor (RD) and a Medical Doctor (MD) shows that they have similarities:

**RD**  I use my medicines, and you employ yours; we are both doctors, and doctors are not deceivers. When a patient dies, you do not give up trust in your medicine, neither do I when rain fails.

**MD**  Could you make it rain on one spot and not on another?

**RD**  I would not think of trying. I like to see the whole country green and all the people glad; the women clapping their hands, and giving me their ornaments for thankfulness, and lullilooing for joy.

*MD*    I think you deceive both them and yourself.

*RD*    Well, then, there is a pair of us (meaning both are rogues).

Taken from Max Gluckman, ed., Introduction to *The Allocation of Responsibility*, pp. xvii–xix. Manchester: Manchester Univ. Press, 1972.

## Notes

1. See, for example, Munasingh (1978).
2. See World Bank (1999). Each year the WDR has a different theme. See also UNDP (1993, 1994, 1995).
3. Some years ago in Papua New Guinea World Bank staff were called "loan salesmen," and even though lending may be going out of style in the aid business, salesmanship is not.
4. See Cochrane (2008).
5. See Williams et al. (1995), Penna and MacFarlane (2012).
6. See Cochrane and Noronha (1973) and Cochrane (1974).
7. See Hicks (1960).
8. In Cochrane and Noronha (1973), this sort of development was not anticipated because at the time of that two-year survey the World Bank had no environmental staff. That changed when Dr. Jim Lee was appointed as its Environmental Advisor. See Goodland (1999).
9. See the discussion on disciplines and fields of activity in Waldo (1965), Dahl (1947).
10. The course outline had work from a US anthropologist Bill Stuart from the University of Maryland as well as work from Rio Tinto's anthropologists Carolyn McCommon and Ramanie Kunanayagam and the author contributed a piece on "culture shock."
11. See Cleveland and Mangone (1957).
12. See Rio Tinto (2012).
13. See Colombia Center on Sustainable Development (2016).
14. When the company had strong community resources the consulting firms had to use Rio Tinto thinking but when Rio Tinto was weak the consulting companies began to try out their own ideas and approaches.
15. Before 1987, the top accountancy firms were referred to as the Big Eight. They were Deloitte Haskins & Sells, Arthur Andersen, Touche Ross, Price Waterhouse, Coopers & Lybrand, Peat Marwick Mitchell, Arthur Young & Co. and Ernst & Whinney. After being involved in the ENRON scandal in the USA Arthur Anderson dropped out of the Big Five.
16. In Mozambique Rio Tinto did not have heavyweight expertise informing the company about the right way to proceed and the results were not good. See Lillywhite et al. (2015).

# Bibliography

Cleveland, Harlan, and Gerard Mangone. 1957. *The Art of Overseasmanship.* Syracuse, NY: Syracuse University Press.

Cochrane, Glynn. 1974. What Can Anthropology Do for Development. *Finance and Development* 11(2): 20.

———. 2008. *Festival Elephants and the Myth of Global Poverty.* New York: Pearson.

Cochrane, Glynn, and Raymond Noronha. 1973. *The Use of Anthropology in Project Operations of the World Bank Group.* Washington, DC: Central Projects, The World Bank.

Colombia Center on Sustainable Development. 2016. *Mapping Mining to the Sustainable Development Goals: A Preliminary Atlas.* New York: Colombia Center on Sustainable Development.

Dahl, R.A. 1947. The Science of Public Administration: Three Problems. *Public Administration Review* 7: 1–11.

Goodland, Robert J. 1999. *Social & Environmental Assessment to Promote Sustainability: An Informal View from the World Bank.* Glasgow: International Association of Impact Assessment.

Hicks, Ursula. 1960. *Development from Below.* Oxford: The Clarendon Press.

Lillywhite, S., D. Kemp, and K. Sturman. 2015. *Mining, Resettlement and Lost Livelihoods: Listening to the Voices of Resettled Communities in Mualadzi, Mozambique.* Melbourne: Oxfam.

Munasingh, M. 1978. *The Leisure Costs of Electricity Light Failure in Developing Countries.* World Bank Paper No. 285, World Bank Staff Working Papers, June. Washington, DC: World Bank.

Penna, Anna-Marie, and Martin MacFarlane. 2012. *Parthenium Incident in the Pilbara, Western Australia: How Is This "A Good News Story"?* Perth: Kellogg Brown & Root Pty Ltd (KBR).

Rio Tinto. 2012. *Communities Standard 2012,* 15–17. London: Rio Tinto.

UNDP. 1993. *Human Development Report 1993.* New York: Oxford University.

———. 1994. *Human Development Report 1994.* New York: Oxford University.

———. 1995. *Human Development Report 1995.* New York: Oxford University.

Waldo, Dwight. 1965. The Administrative State Revisited. *Public Administration Review* 25(1): 5–30.

Williams, K., I. Parer, B. Coman, J. Burley, and M. Braysher. 1995. *Managing Vertebrate Pests: Rabbits.* Canberra: Australian Government Publishing Service, Bureau of Resource Sciences and CSIRO Division of Wildlife and Ecology.

World Bank. 1999. *The World Development Report.* New York: Oxford University Press for the World Bank.

# Appendix A: Environmental Protection After Bougainville[1]

Today, environmental understanding and government regulation have moved on and it is highly unlikely that regulators and investors would approve a mine which depended on riverine tailings disposal. Regulation of mining is an area where the World Bank's Mining Division has provided valuable technical assistance to countries such as Papua New Guinea (PNG), and as a result the PNG authorities and the authorities in other countries have a much stronger capacity to monitor the effects of mining on the environment and to demand remediation when and where necessary or to withdraw the company's mining permit.[2] The history of Freeport's mine in Indonesia which uses riverine tailings disposal shows what can happen when a well-resourced government and a mining company work together to understand and minimize adverse environmental impacts.

Forty years ago riverine tailings disposal was a risk that both the government of PNG and the mining companies were willing to take in order to generate tax and royalty revenues and to provide jobs. Rather than going ahead with mining others would doubtless have voted to leave the minerals in the ground as was the case with Canford. At the time when Bougainville Copper Limited (BCL) and Ok Tedi were permitted, NGOs were not as influential in the decisions that were made about tailings disposal as they would be today. Moreover, the monetary fines and compensation now levied on environmental offenders provide a brake on risk taking.

© The Author(s) 2017                                                       207
G. Cochrane, *Anthropology in the Mining Industry*,
DOI 10.1007/978-3-319-50310-3

Contemporary attitudes and the record suggest that BCL and Ok Tedi should not be used to characterize all mining because BCL and Ok Tedi risks were unique both historically and geographically and because not all mining produces toxic waste that requires impoundment. For BCL and Ok Tedi riverine disposal of tailings was seen as the best option since mining was located in a remote mountainous seismically unstable terrain—the famous Pacific Rim of Fire—which made reliance on dams or on transportation of tailings by long pipelines extremely risky. Building massive tailings dams at BCL and Ok Tedi and then having them catastrophically fail in an earthquake would have been a worse outcome than riverine tailings disposal.[3]

## THE SAMARCO IRON ORE MINE (MINAS GERAIS STATE), BRAZIL

The Samarco tailings dam failed in November 2016 killing 14 employees and contractors and 5 community residents from a small village downstream, which was partially buried in tailings and largely destroyed. The dam is in the upper reaches of the Rio Doce. Muddy water took a few days to travel all the way to the mouth of the Rio Doce, 650 kilometer downstream, and created a plume offshore. There were large fish kills all along the river caused by extremely high turbidity and low oxygen levels in the muddy water. The oxygen levels rose again within several days but the turbidity levels have taken months to gradually drop down and approach normal levels. Fish began returning after a few weeks (there is no public domain data on pre- and post-dam break fish populations yet). Villages and towns that relied on the river for water supplies were provided with alternative potable sources by the company until turbidity dropped enough for water treatment plants to cope again. All the environmental impacts were physical and not chemical, that is, deposition of smothering material on river banks and in the channel, and high suspended solids (turbidity) in the water reducing light (for algae and plankton). The iron ore tailings are chemically inert and comprise the fine fraction washed from the run-of-mine iron ore and as a result although there was considerable physical damage, there were no ongoing problems with toxicity.

There is an agreement to institute remedial environmental and community works to the value of R\$20 billion (US\$5.5 billion) over 5–10 years. Apart from that, there are court cases for damages and compensation due to impacts, and criminal charges pending Samarco, Vale, and the dam inspection company (but not BHP Billiton) in relation to possible negligence relating to the reason the dam failed. The Brazilian Justice approved on Thursday, 5 May, the agreement signed between Samarco, its parent companies Vale and BHP Billiton, the federal government, the state governments of Minas Gerais and Espírito Santo, and other government entities. Signed on 2 March, the agreement has 41 programs of social, environmental, and economic rehabilitation of the areas impacted by the Fundão Dam burst. "The approval is the recognition by the courts of our commitment to the recovery of the impacted areas. This agreement represents a breakthrough in the way of dealing with major issues involving the public power and the private sector, in defense of the interests of society and of populations impacted by accidents like this," said Roberto Carvalho, Samarco's CEO.[4]

## Papua New Guinea

In 1999, BHP reported that 90 million tons of tailings was annually discharged into the river for more than ten years and destroyed downstream villages, agriculture, and fisheries. Mine wastes were deposited along 1000 kilometers (620 miles) of the Ok Tedi and the Fly River below its confluence with the Ok Tedi, and over an area of 100 square kilometers (39 square miles). BHP's CEO, Paul Anderson, said that the Ok Tedi Mine was "not compatible with our environmental values and the company should never have become involved." As of 2006[update], mine operators continued to discharge 80 million tons of tailings, overburden, and mine-induced erosion into the river system each year. About 1588 square kilometers (613 square miles) of forest has died or is under stress. As many as 3000 square kilometers (1200 square miles) may eventually be harmed, an area equal to the US state of Rhode Island or the Danish island of Funen. The mining operation Ok Tedi discharged tailings into the river because there was no impoundment.[5]

The mine is still in operation and waste continues to flow into the river system. BHP was granted legal indemnity from future mine-related damages. The Ok Tedi Mine was scheduled to close in 2013. In 2013, the PNG government seized 100 percent ownership of Ok Tedi Mine and repealed laws that would allow people to sue mining giant BHP Billiton over environmental damage. Ok Tedi Mining Limited launched the OT2025 project that was focused on transitioning the business to a smaller operation in preparation for Mine Life Extension.[6]

## PHILIPPINES

The Marcopper Mining Disaster occurred on 24 March, 1996, on the Philippine island of Marinduque, a province of the Philippines located in the Mimaropa region in Luzon. It remains one of the largest mining disasters in Philippine history. A fracture in the drainage tunnel of a large pit containing leftover mine tailings led to a discharge of toxic mine waste into the Makulapnit-Boac river system and caused flash floods in areas along the river. One village, Barangay Hinapulan, was buried in 6 feet of muddy floodwater, causing the displacement of 400 families. Twenty other villages had to be evacuated. Drinking water was contaminated killing fish and freshwater shrimp. Large animals such as cows, pigs, and sheep were overcome and killed. The flooding caused the destruction of crops and irrigation channels. Following the disaster, the Boac river was declared unusable.[7]

The toxic spill caused flash floods that isolated five villages, with populations of 4400 people each, along the far side of the Boac river. One village, Barangay Hinapulan, was buried under 6 feet of muddy floodwater, causing 400 families to flee to higher grounds. Sources of drinking water were contaminated with toxins. Fish, freshwater shrimp, and pigs were killed outright. Helicopters had to fly in food, water, and medical supplies to the isolated villages. The inhabitants of 20 of the 60 villages in the province were told to evacuate.

## SPAIN

The Doñana Disaster, also known as the Aznalcóllar Disaster or Guadiamar Disaster, was an industrial accident in Andalusia, southern Spain. On 25 April 1998, a holding dam burst at the Los Frailes mine, near Aznalcóllar, Seville Province, releasing 4–5 million cubic meters of mine tailings. The acidic tailings, which contained dangerous levels of several heavy metals, quickly reached the nearby River Agrio, and then its effluent reached the River Guadiamar, traveling about 40 kilometers along these waterways before they could be stopped. The Guadiamar is the main water source for the Doñana National Park, a UNESCO World Heritage Site and one of the largest national parks in Europe. The cleanup operation took three years, at an estimated cost of €240 million. The Los Frailes mine is owned by Boliden-Apirsa (formerly Andaluza de Piritas, SA), the Spanish subsidiary of Boliden, and produces about 125,000 tons of zinc and 2.9 million ounces of silver per year.[8]

## HUNGARY

European environmental groups had long fretted about an aging industrial sludge pond near Ajka, Hungary, containing caustic waste from the process that converts bauxite to aluminum. That all changed around lunchtime on 4 October, when a corner of the sludge reservoir gave way after weeks of heavy rains, letting loose a tidal wave of thick red sludge that oozed its way over garden fences, onto front porches, and into middle-class living rooms in Kolontar and Devecser, where lunches sat waiting on tables. Ten people were killed by the muck—most from drowning—and more than 100 had chemical burns from the highly alkaline mud that were serious enough to require hospitalization.

The Hungarian government reacted swiftly to the disaster, deploying soldiers, firefighters, police, rescue workers, and scientists. They took samples of the mud to evaluate its chemical content, since sludge from bauxite processing can sometimes contain heavy metals and traces of radioactivity as well. They poured acid into the sludge-clogged local rivers—where all life perished immediately—to neutralize the alkaline spill and to prevent its caustic material from reaching the downstream Danube. With shovels and backhoes, teams of workers began removing the veil of sludge from Kolontar, hosing down homes, yards, streets—and even goo-covered animals. The prime minister initiated a criminal investigation, and police briefly imprisoned the reservoir's owner, although he was later released for lack of evidence. After the leading edge of the spill reached the Danube, it was so diluted that the pH remained fine for aquatic life—if not entirely normal. Even in the villages, levels of heavy metals in the sludge were safe, scientists said.[9]

# Appendix B: Data for Community Understanding

Those doing a baseline should make an inventory of everything that has been written about the community involved either in the form of scholarly articles, academic theses, government reports, and aid agency documents. Existing reports and the literature should be examined to see if it can add to existing understanding of the primary factors of community experience related to agriculture, hunting/gathering, and fishing—and should develop quantitative material to support what is said. Estimates as to the extent to which the factors of production are transacted by market forces—the impact of monetization, the elasticity of demand for money—and the continuing role of traditional forms of wealth should be made. Information should also be obtained on savings and investment as well as the circulation of money.

The baseline survey should serve strategic purposes by confirming ways in which the company can work with the community as well as areas where progress will require close or closer cooperation and coordination with government and other parties. From opening to closing it will be advantageous to anticipate the need for coordination and cooperation with local and central government as well as major aid agencies and NGO organizations, and as a result their strengths and weaknesses need to be assessed. Companies should become familiar with the main lines of regional and national development plans of the major public and private sector players as well as relevant policy documents. This information will help to determine the shape of what it is that the company should do and those areas where partnerships may make sense.

© The Author(s) 2017
G. Cochrane, *Anthropology in the Mining Industry*,
DOI 10.1007/978-3-319-50310-3

Public and academic understanding of community relations still suffers from the fact that there are few well-informed studies of miners or NGOs by insiders, too few studies of indigenous peoples and mining, and too few detailed studies of indigenous land tenure or local systems of decision-making related to mining. PNG, Australia, South Africa, and Canada are relatively well represented in the literature because in those countries the regulatory regime has required sensitivity to quite narrow and specialized social concerns.[10] But it is still the case that there are too few studies of mining companies as a community and too few attempts to illustrate the value that can be had when companies are examined with the same professional care that is given to traditional people.[11]

Few mining companies have a well-thought-through strategy to encourage graduate students to undertake dissertation research or to facilitate the work of established academics and other researchers in areas of mutual interest. The need for the results of research information will continue throughout the life of a mine and as a result the baseline ought to examine how to access institutional or other sources. Mining companies spend large sums each year asking consulting firms to collect information without having a clear idea about what is already available or in progress as a book or article or thesis from libraries, universities, and individual researchers. Environmental and Social Impact assessments written by consulting firms seldom contain bibliographic material, and as a consequence it is difficult to assess just how much original work has been undertaken.

The seasonal organization of production as well as the distribution and consumption of production at individual and family levels should be assessed. Where possible information should be collected on time allocation and preferences for men and women, young and old, to assist with estimations of the amount of time and effort, by age and by gender, spent on traditional and "modern" pursuits. The work should provide a clearer understanding of the division of labor, specialization, resource distribution, the economic calendar; hunting, gathering and cultivation; and ways of organizing for economic work.

Different decision-making contexts—ritual, economic, communal welfare, and so on—should be identified as well as the different forms and exercises of authority both ascribed and achieved in situations involving minor infractions of traditional sanctions along with instances of deviance. A baseline should also deal with general processes of social control as well as with the role of authority in dispute resolution situations and the use of authority where government and government decisions are involved.

The strength of well-designed questionnaire surveys lies in their ability to answer specific questions and to answer these authoritatively. But expectations need to be realistic since the quality and reliability of quantitative data roughly correlate with a particular country's level of socioeconomic development: very poor countries have very poorly developed data collection systems, and what is collected is usually not very accurate. Figures for the number of trees planted or land cleared and placed under cultivation are often produced by people who have no idea of what an acre or hectare actually is—kilos and tons of production may be counted by people who have only a vague idea of what a kilo or a ton actually is. In parts of West Africa, for example, weight is equated with truck load. Yields are hard to estimate from farmers' own records and data: harvesting may be spread over some months so that totals are neither noticed nor recorded; harvesting may fail to record the amount used for family consumption and for barter, recording only cash sales in local markets; figures might be falsified to avoid taxes or rent increases. Markets may be officially controlled, resulting in extensive smuggling and black-market trading.

## NOTES

1. How to safely dispose of tailings was the issue in all these major environmental disasters. During mining the ore-bearing rock is ground up into powder and mixed with water and chemicals into slurry. Compressed air is pumped through the mixture causing the copper to hold to the bubbles. These bubbles are then skimmed off the surface and the minerals can be used. The rest of the mixture or tailings are waste. Historically, tailings were disposed of however was convenient, such as in downstream running water, but because of concerns about these sediments in the water and other issues, tailings ponds came into use. However, in areas of the world where there was always a danger of strong seismic activity, or unusually high levels of rainfall, dam construction was not always possible. It was estimated in 2000 that there were about 3500 active tailings impoundments in the world. Because the greatest danger of tailings ponds is dam failure, great care must be taken to ensure sound construction and regular inspection; to detect the possibility of early dam failure is essential and each dam is supposed to be inspected once a year by a qualified engineer. Despite these efforts there have been dam failures in developed as well as developing countries.

2. The World Bank's Mining Division had helped a number of countries develop laws and regulations. Mining companies seconded staff to work in this division of the World Bank.
3. I am indebted to my former colleague Alan Irving for advice on the tailings issue.
4. See, http://www.samarco.com/en/comunicados/http://www.samarco.com/wp-content/uploads/2016/05/Justice-endorses-agrement-for-social-and-environmental-rehabilitation-of-Minas-Gerais-and-Espirito-Santo.pdf.
5. When the author was living in Port Moresby in the 1980s the national newspaper the *Post Courier* had stories about BHP having capsized a barge on the Fly Rover with the loss of 70 drums of cyanide.
6. See, *Ok Tedi Riverine Disposal Case Study*, MMSD, 2002.
7. See Hamilton-Paterson (1997).
8. See Achterberg et al. (1996).
9. See Rosenthal (2010).
10. See Perry (1996).
11. Social scientists who wish to claim to be professional in their work cannot merely apply their scientific training to those organizations and institutions that they approve of, and then use selectivity and bias in examining the "bad" organizations. Surely there must also be an obligation to apply the same scientific standards to mining companies whose activities may not be approved? See Cochrane comment on Roger Keesing's approach to colonial administrators in *OCEANIA*, Vol. 49, Issue 3, March, 1979, p. 235.

## BIBLIOGRAPHY

Achterberg, Elizabeth, et al. 1996. Impact of Los Frailes Mine Spill on Riverine, Estuarine and Coastal Waters in Southern Spain. *Water Resources* 33(16): 3327–3394.

Hamilton-Paterson, James. 1997. A Watery Grave. *Outside Magazine*, January.

Perry, Richard. 1996. *From Time Immemorial: Indigenous Peoples and State Systems*. Austin: University of Texas Press.

Rosenthal, Elizabeth. 2010. *Environment 360*. New Haven, CT: Yale University School of Forestry and Environmental Science.

# BIBLIOGRAPHY

Ali, Saleem H. 2009. *Treasures of the Earth: Need, Greed and a Sustainable Future.* New Haven: Yale University Press.

Alston, Philip, and Gerard Quinn. 1987. The Nature and Scope of States Parties' Obligations Under the International Covenant on Economic, Social and Cultural Rights. *Human Rights Quarterly* 9(2): 156–229.

Angrosino, Michael V., ed. 1976. *Do Applied Anthropologists Apply Anthropology?* Athens: University of Georgia Press.

Becker, Wilfrid. 1996. *Small is Stupid: Blowing the Whistle on the Greens.* London: Duckworth.

Bellamy, John. 1979. *The Tudor Law of Treason.* London: Routledge.

Belshaw, Cyril. 1976. *The Sorcerer's Apprentice: An Anthropology of Public Policy.* New York: Pergamon.

Bhuta, N. 1998. Mabo, Wik and the Art of Paradigm Management. *Melbourne University Law Review* 22: 24–41.

Biersack, Aletta. 1999. The Mount Kare Python and His Gold: Totemism and Ecology in the Papua New Guinea Highlands. *American Anthropologist* 101(1): 68–87.

Bird, Deborah Rose. 1996. Land Rights and Deep Colonising: The Erasure of Women. *Aboriginal Law Bulletin* 3(85): 6–13.

Bohannon, Paul. 1957. *Justice and Judgement Among the Tiv.* Oxford: Oxford University Press.

Bort, J.S. 1981. *An Environmental Reconnaissance of the Cerro Colorado Concession Area, Empresa de Cobre Cerro Colorado,* S.A. Panama City.

Brereton, David, and Bruce Harvey. 2005. Emerging Models of Community Engagement in the Australian Minerals Industry. A paper presented at the International Conference on Engaging Communities, An Initiative of the United Nations and the Queensland Government, Brisbane, August 14–17.

© The Author(s) 2017                                                                                        217
G. Cochrane, *Anthropology in the Mining Industry,*
DOI 10.1007/978-3-319-50310-3

Brokensha, David. 1966. *Applied Anthropology in English-Speaking Africa.* Ithaca, NY: Society for Applied Anthropology, Monograph 8.

Brokensha, David, Michael Warren, and Oswald Werner. 1980. *Indigenous Knowledge Systems and Development.* Lanham, MD: University Press of America.

Brosius, Peter J. 1999. Analysis and Interventions: Anthropological Engagement with Environmentalism. *Current Anthropology* 40(3): 277–310.

Brysk, Alison. 1996. The Internationalization of Indian Rights. *Latin American Perspectives* 2(3), Ethnicity and Class in Latin America (Spring).

Buchanan, James, and Gordon Tullock. 1962. *The Calculus of Consent.* Ann Arbor: University of Michigan Press.

Callahan, Michael D. 1999. *Mandates and Empire: The League of Nations and Africa, 1914–1931.* Brighton: Sussex Academic Press.

Campisi, Jack. 1976. The New York-Oneida Treaty of 1795: A Finding of Fact. *American Indian Law Review* 71–82.

Carstairs Morris, G. 1958. *The Twice Born: A Study of a Community of High-Caste Hindus.* Bloomington, IN: Indiana University Press.

Cattano, Ben. 2009. *The New Politics of Natural Resources: Time for Extractive Industries to Address Above-Ground Performance.* London: Environmental Resources Management.

Chambers, Robert. 1981. Rapid Rural Appraisal: Rationale and Repertoire. *Public Administration and Development* 1: 95–106.

Chatham House. 2013. *Revisiting Approaches to Community Relations in Extractive Industries: Old problems, New Avenues?* London: Chatham House.

Cleveland, Harlan, and Gerard Mangone. 1957. *The Art of Overseasmanship.* Syracuse, NY: Syracuse University Press.

Coates, Ken, ed. 2013. *From Aspiration to Inspiration: UNDRIP Finding Deep Traction in Indigenous Communities.* The Center for International Governance Innovation(CIGI).https://www.cigionline.org/blogs/aspiration-inspiration-undrip-finding-deep-traction-indigenous-communities/

Cobo, José Martinez. 1987. *Study of the Problem of Discrimination Against Indigenous Populations.* UN Document E/CN.4/Sub.2/1986/7.

Cochrane, Glynn. 1969a. Strategy in Community Development. *Journal of Developing Areas* 8: 5–12.

———. 1969b. Choice of Residence in the Solomons and a Focal Land Model. *Journal of the Polynesian Society* 78(3): 330–343.

———. 1969c. The Administration of Wagina Resettlement Scheme. *Human Organization* 29(Summer): 123–132.

———. 1970. *Big Men and Cargo Cults.* Oxford: Oxford University Press.

———. 1971a. *Development Anthropology.* New York: Oxford University Press.

———. 1971b. Juristic Persons, Group and Individual Land Tenure: A Rejoinder to Goodenough. *American Anthropologist* 73(5): 1152–1155.

———. 1974a. What Can Anthropology Do for Development. *Finance and Development* 11(2): 20.

———. 1974b. Land Alienation: The Case for Traditionalists. *Oceania* 45(2): 124–131.

———. 1976. *What We Can Do for Each Other: An Interdisciplinary Approach to Development Anthropology*. Amsterdam: B.R. Gruner.

———. 1979. *The Cultural Appraisal of Development Projects*. New York: Praeger.

———. 1982. Review of David Brokensha, Michael Warren, and Oswald Werner, *Indigenous Knowledge Systems and Development*. Lanham, MD: University Press of America, 1980, in *American Anthropologist* 84(2).

———. 2008. *Festival Elephants and the Myth of Global Poverty*. New York: Pearson.

Cochrane, Glynn, and Raymond Noronha. 1973. *The Use of Anthropology in Project Operations of the World Bank Group*. Washington, DC: The World Bank, Central Projects.

Colombia Center on Sustainable Development. 2016. *Mapping Mining to the Sustainable Development Goals: A Preliminary Atlas*. New York: Colombia Center on Sustainable Development.

Colson, E. 1958. *Marriage and Family Among the Plateau Tonga of Northern Rhodesia*. Manchester: Manchester University Press.

Conner, Kathleen R. 1991. A Historical Comparison of Resource-Based Theory and Five Schools of Thought within Industrial Organization Economics: Do We Have a New Theory of the Firm. *Journal of Management* 17(1): 121–154.

Culhane, Dara. 1998. *The Pleasure of the Crown: Anthropology, Law, and First Nations*. Burnaby, BC: Talon Books.

Curely, Edward, ed. 1994. *Thomas Hobbes, Leviathan*. Indianapolis: Hackett.

Dahl, R.A. 1947. The Science of Public Administration: Three Problems. *Public Administration Review* 7: 1–11.

Davidson, Jeffrey. 1993. The Transformation and Successful Development of Small-Scale Mining Enterprises in Developing Countries. *Natural Resources Forum* 17(4): 315–326.

Davis, Shelton. 1977. *Victims of the Miracle: Development and the Indians of Brazil*. Cambridge: Cambridge University Press.

Davis, R., and D. Franks. 2011. The Cost of Conflict with Local Communities in Extractive Industry. In *Proceedings of the First International Seminar on Social Responsibility in Mining*, eds. D. Brereton, et al., ICMM, Santiago, October 19–21.

Demtchev, Nikolay. 2004. Corporate Social Performance as a Business Strategy. *Journal of Business Ethics* 55(4): 395–410.

Denoon, Donald. 2000. *Getting Under the Skin: The Bougainville Copper Agreement and the Creation of the Panguna Mine*. Melbourne: Melbourne University Press.

Derrett, J.D.M. 1978. *Essays in Classical and Modern Hindu Law*. Leiden: Brill.

De Soto, Hernando. 2000. *The Mystery of Capital, Why Capital Triumphs in the West and Fails Everywhere Else*. London: Bantam Press.

Dostal, W., ed. 1972. *The Situation of the Indian in South America*. Geneva: World Council of Churches.

Dreger, Alice. 2011. Darkness's Descent on the American Anthropological Association: A Cautionary Tale. *Human Nature* 22: 225–246.

Dundes, Alison Renteln. 1990. *International Human Rights: Universalism versus Relativism*. London: Sage.

———. 2004. *The Cultural Defense*. New York: Oxford University Press.

Edwards, Michael. 2000. *NGO Rights and Responsibilities, A New Deal for Global Governance*. London: The Foreign Policy Center.

Edwards, Michael, and David Hulme. 1992. *Making a Difference: NGOs and Development in a Changing World*. London: Earthscan.

Eliade, Mercia. 1962. *The Forge and the Crucible*. London: Rider.

Elkington, J. 1997. *Cannibals with Forks: The Triple Bottom Line of 21st Century Business*. Oxford: Capston.

Epstein, A.L. 1969. *Matupit, Land, Politics and Change Among the Tolai of New Britain*. Berkeley: University of California Press.

Erasmus, Charles. 1961. *Man Takes Control*. Minneapolis, MN: University of Minneapolis Press.

Esty, Daniel, and Maria Ivanova. 2001. Making International Environmental Efforts Work: The Case for a Global Environmental Organization. Paper prepared for an open meeting of the Global Environmental Change Research Community, Rio de Janeiro.

Evans-Pritchard, E.E. 1940. *The Nuer: A Description of the Modes of Livelihood and Political Institutions of a Nilotic People*. Oxford: The Clarendon Press.

———. 1965. *Theories of Primitive Religion*. Oxford: The Clarendon Press.

Fassin, Yves. 2008. SMEs and the Fallacy of Formalising CSR. *Business Ethics: A European Review* 17(4): 364–378. doi:10.1111/j.1467-8608.2008.00540.x.

———. 2009. The Stakeholder Model Refined. *Journal of Business Ethics* 84(1): 113–135.

———. 2010. A Dynamic Perspective in Freeman's Stakeholder Model. *Journal of Business Ethics* 96: 39–49.

Filer, Colin. 1990. The Bougainville Rebellion, The Mining Industry and the Process of Social Disintegration in Papua New Guinea. *Canberra Anthropology* 13(1): 1–39.

Flynn, Sharon, and Liz Vergara. 2015. *Land Access and Resettlement Planning at La Granja*. CSRM Occasional Papers: Mining Induced Resettlement Series, Centre for Social Responsibility in Mining, Queensland University, St. Lucia, Brisbane.

Forde, Daryll. 1934. *Habitat, Economy and Society*. London: Methuen.

Forni, N. 2002. *Land Tenure Policies in the Near East*. Rome: FAO.

Forster, George M. 1969. *Applied Anthropology*. Boston: Little Brown.

Fortes, Meyer. 1963. Ritual and Office in Tribal Society. In *Essays on the Ritual of Social Relations*, ed. Max Gluckman. Manchester: Manchester University Press.

Foster, George M. 1965. Peasant Society and the Image of Limited Good. *American Anthropologist* 67(2): 293–315.

Foucault, Michele. 1995. *Discipline and Punish: The Birth of the Prison*. New York: Vintage Books.

Frake, Charles O. 1964. How to Ask for a Drink in Subanun. *American Anthropologist* 66(6), Pt 2: 127–130.

Franks, Daniel, Courtney Fidler, David Brereton, Frank Vanclay, and Phil Clark. 2009. *Leading Practice Strategies for Addressing the Social Impacts of Resource Developments*. Brisbane: Center for Social Responsibility in Mining, Sustainable Minerals Institute, The University of Queensland & Department of Employment, Economic Development and Innovation, Queensland Government.

Freeman, R., Jeffrey Harrison, and Andrew C. Wicks. 2007. *Managing for Stakeholders: Survival, Reputation and Success*. New Haven: Yale University Press.

Gauthier, David. 1988. *Morals by Agreement*. Oxford: Blackwell.

Geertz, Clifford. 1984. Anti Anti-Relativism. *American Anthropologist* 86(2): 263–278.

Gjording, C. 1981. *The Cerro Colorado Copper Project and the Guaymi Indians of Panama*. Occasional Papers No. 3. Cambridge, MA: Cultural Survival.

Godoy, Ricardo. 1985. Mining: Anthropological Perspectives. *American Review of Anthropology* 14: 199–217.

Goethe. 1867. *Faust*, Part 1: 1112, English translation by John Wynniatt Grant.

Goodland, Robert J. 1999. *Social & Environmental Assessment to Promote Sustainability: An Informal View from the World Bank*. Glasgow: International Association of Impact Assessment.

Gordon, Robert. 1978. The Celebration of Ethnicity: A Tribal Fight in a Namibian Mine Compound. In *Ethnicity in Modern Africa*, ed. Brian du Toit. Boulder, CO: Westview Special Studies on Africa.

Graicunas, V.A. 1937. Relationships in Organizations. In *Papers in the Science of Administration*, ed. Luther Gulick and F. Urwick Lyndall. New York: Colombia University Institute of Public Management.

Greene, Graham. 1984. *Getting to Know the General*. London: Bodley Head.

Grillo, Ralph, and Alan Rew, eds. 1985. *Social Anthropology and Development Policy*. ASA Monograph No. 23. London: Tavistock Publications.

Hage, Per. 2004. East Papuan Kinship Systems: Bougainville. *Oceania* 75(2): 109–124.

Hageboeck, Molly, Glynn Cochrane, Lawrence Cooley, and Gerald Hursh-Cēsar. 1979. *The Manager's Guide to Data Collection*. Washington, DC: Agency for International Development.

Hall, Gillette, and Harry Anthony Patrinos. 2013. *Indigenous Peoples, Poverty and Human Development in Latin America*. New York: Palgrave Macmillan.

Hamalainen, Pekka. 2008. *The Comanche Empire*. New Haven: Yale University Press.

Hanai, Maria. 2000. Formal and Garimpo Gold Mining and the Environment in Brazil. In *Mining and the Environment: IDRC Case Studies from the Americas*, ed. A. Warhurst. Ottawa: Government of Canada.

Harvey, W.B. 1966. *Law and Social Change in Ghana*. Princeton, NJ: Princeton University Press.

Heider, Karl. 1970. *The Dugum Dani: A Papuan Culture in the Highlands of West New Guinea*. Chicago: Aldine Publishing.

Henriksen, John B. 2008. *Key Principles in Implementing ILO Convention No. 169*. Geneva: ILO.

Herskovits, Melville. 1952. Economics and Anthropology, A Rejoinder. *Journal of Political Economy* XLIX(2): 269–278.

Hicks, Ursula. 1964. *Development from Below*. Oxford: The Clarendon Press.

Hoben, Alan. 1982. Anthropologists and Development. *Annual Review of Anthropology*, American Anthropologist, 11: 349–375.

Hoessle, Ulrike. 2014. The Contribution of the UN Global Compact towards the Compliance of International Regimes: A Comparative Study of Businesses from the USA, Mozambique, United Arab Emirates and Germany. *Journal of Corporate Citizenship* 53: 27–60.

Holcomb, Sarah. 2004. Traditional Owners and 'Community Country' *Anangu*: Distinctions and Dilemmas. *Australian Aboriginal Studies* 2: 64–71.

Holme, Richard, and Phil Watts. 2000. *Corporate Social Responsibility: Making Good Business Sense*. Geneva: World Business Council for Sustainable Development.

Hooker, M.B. 1978. *Adat Law in Modern Indonesia*. Oxford: Oxford University Press.

Hyndman, David. 1987. Mining, Modernization, and Movements of Social Protest in Papua New Guinea. *Social Analysis* 21: 20–38.

Hyndman, David C. 1997. The Archaeology and Anthropology of Mining: Social Approaches to an Industrial Past. *Current Anthropology* 38: 20–32.

ICMM. 2014. *The Role of Mining in National Economies*. 2nd ed. London: ICMM.

Inglis, Julian. 1993. *Traditional Ecological Knowledge: Concepts and Cases International Program on Traditional Ecological Knowledge*. Ottawa, Canada: International Development Research Center.

International Federation of Chemical Engineering and Mine Workers. 1997. *Rio Tinto: Tainted Titan, The Stakeholders Report*. Brussels: International Federation of Chemical Engineers and Mine Workers.

International Institute for Environment and Sustainable Development. 2002. *Breaking New Ground: Mining, Minerals and Sustainable Development*. London: IIED.

International Union for the Conservation of Nature. n.d. *World Directory of National Parks.* International Union for the Conservation of Nature: Glans, Switzerland.

Jackson, Richard. 1991. Not Without Influence: Villagers, Mining Companies and Governments in Papua New Guinea. In *Mining and Indigenous Peoples in Australasia*, ed. J. Connell and R. Howitt. Sydney: Sydney University Press.

Jaen, Bernardo. 1982. *El Impacto Del Project De Cerro Colorado in El Pueblo Guaymi y Su Futuro.* Centro De Estudio Y Accion Social: Panama.

Jorgenson, Dan. 1997. Who and What is a Landowner? Mythology and Marking the Ground in a Papua New Guinea Mining Project. *Anthropological Forum* 7(4): 599–627.

Joyce, Susan, and Ian Thompson. 2000, February. Earning a Social License to Operate: Social Acceptability and Resource Development in Latin America. *The Canadian Mining and Metallurgical Bulletin* 93(1037): 49–53.

Keesing, Roger. 1989. *Creating the Past: Custom and Identity in the Contemporary Pacific.* Honolulu: University of Hawaii Press.

Kemp, D., and J.R. Owen. 2013. Community Relations and Mining: Core to Business but Not 'Core Business'. *Resources Policy* 38(4): 523–531.

Kenan Institute for Ethics. 2012. *The U.N. Guiding Principles on Business and Human Rights: Analysis and Implementation.* Durham, North Carolina: Kenan Institute for Ethics, Duke University.

Kirsch, Stuart. 2014. *Mining Capitalism: The Relationship between Corporations and Their Critics.* Berkeley, CA: University of California Press.

———. 2015. *Reverse Anthropology: Indigenous Analysis of Social and Environmental Relations in New Guinea.* Palo Alto, CA: Stanford University Press.

Knight, F.H. 1941. Anthropology and Economics. *Journal of Political Economy* XLIX(2): 247–268.

Knox, Ronald A. 1950. *Enthusiasm.* New York: Oxford University Press.

Korten, David C. 1986. Introduction. In *Go To The People*, ed. James B. Mayfield. West Hartford, CT: Kumarian Press.

Kuiper, Andrew. 2004. Harnessing Corporate Power: Lessons from the UN Global Compact. *Bulletin on the Development of Federalism* 47: 9–19.

Labonne, Béatrice. 1996. Artisanal Mining: An Economic Stepping Stone for Women. *Natural Resources Forum* 20(2): 117–122.

Langton, Marcia. 2012. Third Boyer Lecture, Melbourne (December).

Laracy, Hugh. 1976. *Marists and Melanesians; A History of the Catholic Missions in the Solomon Islands.* Honolulu: University Press of Hawaii.

Laslett, Peter, ed. 1960. *Locke: Two Treatises of Government.* Cambridge: Cambridge University Press.

Lasslett, Kristian. 2009. *Winning Hearts and Mines: The Bougainville Crisis, 1988–90.* London: Routledge.

Lawoti, Mahendra, and Anup Kumar. 2009. *The Maoist Insurgency in Nepal: Revolution in the Twenty-first Century.* London: Routledge.

Leith, Denise. 2003. *The Politics of Power: Freeport in Suharto's Indonesia.* Honolulu: University of Hawaii Press.

Lillywhite, S., D. Kemp, and K. Sturman. 2015. *Mining, Resettlement and Lost Livelihoods: Listening to the Voices of Resettled Communities in Mualadzi, Mozambique.* Melbourne: Oxfam.

Linton, Ralph. 1943. Nativistic Movements. *American Anthropologist* 45(2): 230–240.

Lofstedt, Ragnar E., and Ortwin Renn. 1997. The Brent Spar Controversy: An Example of Risk Communication Gone Wrong. *Risk Analysis* 17(2): 131–135.

Lunenburg, Fred C. 2010. Managing Change: The Role of the Change Agent. *International Journal of Management, Business and Administration* 13(1): 1–16.

Lynch, Larry. 2014. History of the Crandon Mine Project and Implications of Lessons Learned. SME Wisconsin Annual Conference on Partnering for Sustainable Mining: Hard Rock and Soft Rock Mining in Minnesota and Wisconsin, Eau Claire, Wisconsin.

Macintyre, Martha. 2007. Informed Consent and Mining Projects: A View from Papua New Guinea. *Pacific Affairs* 80(1, Spring): 49–65.

Maitland, F.W. 1908. *The Constitutional History of England: A Course of Lectures.* Cambridge: Cambridge University Press.

Malinowski, Bronisław. 1929. Practical Anthropology. *Africa* 2(1): 22–38.

Masefield, Geoffrey. 1990. Agricultural Extension. In *The Bougainville Crisis*, ed. R.J. May and M. Spriggs. Bathurst: Crawford House Press.

Mayer, J.P. 1955. *Max Weber and German Politics.* London: Faber and Faber.

McIntosh, Malcolm, Deborah Leipziger, Keith Jones, and Gill Coleman. 1998. *Successful Strategies for Responsible Companies.* London: Financial Times Management.

Mealey, George. 1996. *Grasberg—Mining the Richest and Most Remote Deposit of Copper and Gold in the World in the Mountains of Irian Jaya, Indonesia.* Singapore: Freeport-McMoRan Copper & Gold Inc.

Metx, Shannah. 2006. Indigenous People's Right to Free Prior Informed Consent (FPIC) and Project Governance. *Collaborator for Research on Global Projects.* Stanford University.

Migdal, Joel S. 1988. *Strong Societies and Weak States: State-Society Relations and State Capabilities in the Third World.* Princeton, NJ: Princeton University Press.

Moore, M. 1998. Corporate Governance for NGO? *Development in Practice* 8(3): 335–342.

Morrison, John. 2014. *The Social License.* London: Palgrave-Macmillan.

Munasingh, M. 1978. *The Leisure Costs of Electricity Light Failure in Developing Countries.* World Bank Paper No. 285, World Bank Staff Working Papers, June. Washington, DC: World Bank.

Nash, J. 1974. *Matriliny and Modernisation: The Nagovisi of South Bougainville*. New Guinea Research Bulletin No. 55, Port Moresby and Canberra, Australian National University.

Nash, June. 1979. *We Eat the Mines*. New York: Colombia University Press.

Nelson, Jane, and Simon Zadek. 2000. *Partnership Alchemy*. Copenhagen Centre: Copenhagen.

O'Faircheallaigh, C. 2006. Mining Agreements and Aboriginal Economic Development in Australia and Canada. *Journal of Aboriginal Economic Development* 5(1): 74–91.

Ogan, Eugene. 1971. Nasioi Land Tenure: An Extended Case Study. *Oceania* XLII(2): 81–93.

———. 1991. The Cultural Background to the Bougainville Crisis. *Journal de la Société des Océanistes* 92(1): 61–67.

———. 1996. Copra Came Before Copper: The Nasioi of Bougainville and Plantation Agriculture 1902–1964. *Pacific Studies* 19: 31–52.

———. 1999. *The Bougainville Conflict: Perspectives from Nasioi*. Technical Report Discussion Paper, Australian National University, Canberra, March.

Oliver, Douglas. 1955. *A Solomon Island Society: The Siwai of Bougainville*. Cambridge: Harvard University Press.

———. 1991. *Black Islanders, A Personal Perspective of Bougainville 1937–1991*. Honolulu: University of Hawaii Press.

Melchett, P. 1995. Green for Danger. *New Scientist* 148(2010): 50–51.

Pakinkas, L.A., et al. 1993. Community Patterns of Psychiatric Disorders After the Exxon Valdez Oil Spill. *American Journal of Psychiatry* 150(10): 474–478.

Palomka, Peter, ed. 1990. Bougainville: Perspectives on a Crisis. *Canberra Papers on Strategy and Defence*, vol. 66.

Penna, Anna-Marie, and Martin MacFarlane. 2012. *Parthenium Incident in the Pilbara, Western Australia: How is This "A Good News Story"?* Perth: Kellogg Brown & Root.

Penna-Friema, Rodrigo, and Eduardo Bredgo. 2007. The Risks of Commoditising Poverty: Rural Communities, Quilombola Identity and Natural Conservation in Brazil. *Habitas, Goiania*, vol. V, July.

Perham, Margery. 1937. *Native Administration in Nigeria*. Oxford: Oxford University Press.

Perry, Richard. 1996. *From Time Immemorial: Indigenous Peoples and State Systems*. Austin, TX: University of Texas Press.

Peterson, N. 1993. Demand Sharing: Reciprocity and Pressure for Generosity Among Foragers. *American Anthropologist* 95(4): 860–874.

Politakis, George P., and K. Kolben. 2010. Labor Rights as Human Rights. *Virginia Journal of International Law* 50: 1–31.

Polyani, M. 1958. *Personal Knowledge: Toward a Post-critical Philosophy*. Chicago: University of Chicago Press.

Prno, Jason. 2013. An Analysis of Factors Leading to the Establishment of a Social License to Operate in the Mining Industry. *Resources Policy* 38(4): 577–590.

RAID (Rights and Accountability in Development). 2015. *Rethinking the UN Guiding Principles and Company Grievance Mechanisms.* Oxford: RAID.

Rawls, John. 1999. *A Theory of Justice.* Cambridge, MA: Belknap Press of Harvard University.

Rio Tinto. 2012. *Communities Standard 2012.* London: Rio Tinto.

Robertson Smith, William. 1889. *Lectures on the Religion of the Semites. Fundamental Institutions.* First Series. London: Adam & Charles Black.

Rotberg, Robert I. 2003. *When States Fail: Causes and Consequences.* Princeton, NJ: Princeton University Press.

Russell, Peter. 2005. *Recognizing Aboriginal Title: The Mabo Case and Indigenous Resistance to English-Settler Colonialism.* Toronto: University of Toronto Press.

Ruttan, Vernon W. 1986. Assistance to Expand Agricultural Production. *World Development* 4(1): 29–63.

Sainsbury, C., C. Wilkins, D. Haddad, D. Sweeney, et al. 2011. *Generation Next: A Look at Future Greenfield Growth Projects, Citi Investment Analysis.* New York: Citibank.

Salazar, Debra J., and Donald K. Alper, eds. 2001. *Forging Truces in the War in the Woods.* Seattle, Washington: University of Washington Press.

Schluter, Michael. 2012. What Charter for Humanity? Defining the Destination of Development. In *After Capitalism: Rethinking Economic Relationships,* ed. Paul Mills and Michael Schluter. Cambridge: Jubilee Centre.

———. 2016. *Three Relational Concerns about the Sustainable Development Goals.* Sustainable Development Goals: The Missing Dimension, A Relational Thinking Dialogue, Unpublished Discussion Paper, Geneva, April.

Scott, Parry. 1997. *Garimpos, Community and Rio Paracatu.* Unpublished Report.

———. 2011. Families, Nations and Generations in Women's International Migration. *Vibrant, Virtual Brazilian, Anthropology* 8(2): 279–306.

Scudder, Thayer. 2007. Pipe Dreams: Can the Zambezi River Supply the Region's Water Needs? *Cultural Survival Quarterly* 31: 7–31.

Semple, Janet. 1983. *Bentham's Prison: A Study of the Panopticon Penitentiary.* Oxford: The Clarendon Press.

Shokeid, Moshe. 2007. From the Tikopia to Polymorphous Engagements: Ethnographic Writing Under Changing Fieldwork Circumstances. *Social Anthropology/Anthropologie Sociale* 15(3): 305–319.

Simpson, S. Rowton. 1976. *Land, Law and Registration.* Cambridge: Cambridge University Press.

Skalnik, Peter. 1989. Lihir Society on the Eve of Mining Operations: A Long Term Project for Urgent Anthropological Research in Papua New Guinea. *Bulletin of the International Committee on Urgent Anthropological Research,* Nos 32–33, Vienna, UNESCO.

Smillie, Ian. 1995. *The Alms Bazaar, Altruism Under Fire - Non-profit Organizations and International Development.* London: Intermediate Technology Publications.

Spicer, Edward H. 1952. *Human Problems in Technological Change: A Casebook.* New York: Wiley.

Stocking, George W. Jr. 1996. *After Tylor: British Social Anthropology 1888–1951.* London: Athlone Press.

Stockman, Lorne, James Marriott, and Andrew Rowell. 2005. *The Next Gulf: London, Washington and Oil Conflict in Nigeria.* Washington, DC: Constable & Robinson.

Sutton, Peter. 1995. *Country: Aboriginal Boundaries and Land Ownership in Australia.* Aboriginal History Monograph No. 3. Canberra: Australian National University.

Swartz, Spencer. 2010. BP Provides Lessons Learned From Gulf Spill. *The Wall Street Journal,* September 5.

Taussig, M. 1980. *The Devil and Commodity Fetishism in South America.* Chapel Hill: University of North Carolina Press.

Taylor, Frederick W. 1919. *The Principles of Scientific Management.* New York: Harpers.

Tennyson, Ros, and Luke Wilde. 1998. *The Guiding Hand: Brokering Partnerships for Sustainable Development.* London: United Nations Staff College, Prince of Wales Business Leaders Forum.

Tjiho, Job, and Grobler, H. n.d. *Draft History of the Rössing Foundation.* Unpublished Document, Rössing Mine, Swakopmund, Namibia.

Tobin, Brendan. 2014. *Indigenous Peoples, Customary Law and Human Rights—Why Living Law Matters.* London: Routledge.

Toensing, Gale. 2013. Political Party! Celebrating UNDRIP and Indigenous Culture in Montreal. *Indian Country Today.* http://indiancountrytodaymedianetwork. com/2011/09/13/political-party-celebrating-undrip-and-indigenous-culture-montreal-53604/

Tonkinson, Robert. 2007. Aboriginal 'Difference' and 'Autonomy' Then and Now: Four Decades of Change in a Western Desert Society. *Anthropological Forum* 17(1): 41–60.

Trigger, David S. 1997. Mining, Landscape and the Culture of Development Ideology in Australia. *Cultural Geographies* 4(2): 161–180.

Tuminez, Astrid. 2013. *This Land is Our Land: Moro Ancestral Domain and Its Implications for Peace and Development in the Southern Philippines.* Baltimore, MD: Johns Hopkins, School for Advanced International Studies.

UN Department of Economic and Social Affairs. 2005. *An Overview of the Principle of Free, Prior and Informed Consent and Indigenous Peoples in International and Domestic Law and Practices.* New York: UN Department of Economic and Social Affairs.

UNDP. 1993. *Human Development Report 1993.* New York: Oxford University.

————. 1994. *Human Development Report 1994.* New York: Oxford University.

————. 1995. *Human Development Report 1995.* New York: Oxford University.

United Nations. 2011. *Implementing the United Nations "Protect, Respect and Remedy" Framework.* New York and Geneva: United Nations.

Utting, Peter. 1993. *Trees, People and Power: Social Dimensions of Deforestation and Forest Protection in Central America.* London: Earthscan.

Varma, Roli. 2005. The Bhopal Disaster of 1984. *Bulletin of Science, Technology and Society* 25(1): 37–45.

Vaughn, Diane. 1996. *The Challenger Launch Decision: Risky Technology, Culture, and Deviance at NASA.* Chicago: University of Chicago Press.

Wade, Robert. 1987. The Management of Common Property Resources: Finding a Cooperative Solution. *The World Bank Research Observer* 2(2): 129–133.

Waldo, Dwight. 1965. The Administrative State Revisited. *Public Administration Review* 25(1): 5–30.

West, Richard. 1972. *River of Tears: The Rise of the Rio Tinto Zinc Corporation Limited.* London: Earth Island.

Williams, F.E. 1939. The Creed of a Government Anthropologist. Presidential Address Australian and New Zealand Association for the Advancement of Science, Sydney, Australia.

Williams, Oliver F. 2004. The UN Global Compact: The Challenge and the Promise. *Business Ethics Quarterly* 14(4): 755–774.

Williams, K., I. Parer, B. Coman, J. Burley, and M. Braysher. 1995. *Managing Vertebrate Pests: Rabbits.* Canberra: Australian Government Publishing Service, Bureau of Resource Sciences and CSIRO Division of Wildlife and Ecology.

Wilson, Richard. 1997. *Human Rights and Cultural Context: Anthropological Perspectives.* London: Pluto Press.

Wilson, Sir Robert. 2000. *Meeting the Challenge to 21st Century Mining.* Davos: World Economic Forum.

Wittfogel, Karl August. 1956. *The Hydraulic Societies.* Chicago: Chicago University Press.

World Bank. 1999. *The World Development Report.* New York: Oxford University Press for the World Bank.

Worsley, Peter. 1964. *The Trumpet Shall Sound.* London: Methuen.

Wright, Laura, and Jerry P. White. 2012. Developing Oil and Gas Resources On or Near Indigenous Lands in Canada: An Overview of Laws, Treaties, Regulations and Agreements. *The International Indigenous Policy Journal* 3(2), Article 5.

Young, Philip D. 1971. *Ngawbe: Tradition and Change Among the Western Guyamí of Panama.* Chicago: Board of Trustees of the University of Illinois.

Zadek, Simon. 2001. *The Civil Corporation: The New Economy of Corporate Citizenship.* London: Earthscan.

# INDEX

## A

aboriginals (Australia), 48, 50–5,
  58n22, 83, 113, 125–6, 175, 177
  land rights, 175
aid agencies
  absence of community relations in
    statistics, 10n15, 43
  global ambitions of, 196
  poverty projects of, 94, 163
  skill set of employees, 93
  social responsibilities of, 9, 163, 165
  use of RRA, 169
Ajka, Hungary, 211
American Anthropological Association
  1947 Statement on Human
    Rights, 67
  Code of Ethics, 27
Amungme, 31, 144, 145
Andaluza de Piritas, S.A. *See*
  Boliden-Apirsa
Anderson, Paul, 209
Angiabbak, Angus, 144
Anglesey Aluminum Metal (AAM). *See
  also* Rio Tinto

mine closure, 123
outsourced social baseline, 8,
  103–8, 114, 126, 130n5, 138,
  140, 154, 163, 166, 170
Annan, Kofi (UN Secretary
  General), 63
anthropological fieldwork, 91,
  104, 169
  deductive method of, 99
anthropologists, v, 4–6, 27, 32n13,
  39, 53, 67, 72, 104–5, 108,
  130n5, 138, 143–4, 203, 205n10
  collaboration with local mining
    teams, 104
  comparative ethnography
    experience, 108
anthropology. *See* cultural
  anthropology
Armchair Anthropology, 72, 77n25
artisanal mining, 8, 135–59, 165
Association for the Welfare of the
  Mining Community and
  Environment (LKMTL),
  156, 158

© The Author(s) 2017
G. Cochrane, *Anthropology in the Mining Industry*,
DOI 10.1007/978-3-319-50310-3

Printed in Great Britain
by Amazon

29336806R00145